# INFRARED AND RAMAN SPECTROSCOPY

R.L. Laksh

**RAJAT PUBLICATIONS**
*New Delhi*

**Infrared and Raman Spectroscopy**
by *R.L. Laksh*

---

This edition is published by Rajat Publications
4740/23, Ansari Road Daryaganj,
New Delhi - 110002

ISBN: 978-81-7880-068-4



Edition 2025



Printed and bound in India

RAJAT
PUBLICATIONS

4740/23, Ansari Road, Daryaganj,
New Delhi- 110002 (India)
Mob:- +91-9811484973, +91-9811105389
E-mail: rajat_publication@yahoo.com

# Preface

This text deals with the theory instrumentation and application of Infrared and Raman spectroscopy. This text is the clearest fullest and at the same time most succinct of the concepts to these two important spectroscopic techniques. Copious illustrations in line reinforce the student understanding of each topic covered. This text is carefully structured to make these techniques as easy and straightforward as possible. This is organized systematically by introducing ideas in their natural order. Graduate and postgraduate students following the courses in spectroscopy and analytical chemistry will find this book to be most useful.

**—Editor**

# Contents

# 1

# Infrared Spectroscopy

## Introduction

Infrared (IR) spectroscopy is one of the most common spectroscopic techniques used by organic and inorganic chemists. Simply, it is the absorption measurement of different IR frequencies by a sample positioned in the path of an IR beam. The main goal of IR spectroscopic analysis is to determine the chemical functional groups in the sample. Different functional groups absorb characteristic frequencies of IR radiation. Using various sampling accessories, IR spectrometers can accept a wide range of sample types such as gases, liquids, and solids. Thus, IR spectroscopy is an important and popular tool for structural elucidation and compound identification.

Vibrational spectra arise from transitions between vibrational energy levels. These energy levels include bond stretching, bond bending, and other internal motions of the molecules. Some vibrational frequencies identify particular groups in the molecule, such as a C—H bond, a C=C double bond, or phenyl group. Other frequencies characterize conformations such as amide frequencies in proteins or phosphate ester frequencies in nucleic acids. Vibrational spectra occur in the infrared region, which may be considered to include the region from approximately $10^3$ nm to $10^5$ nm. The traditional units used by infrared spectroscopists are either microns (μm) or wave-numbers ($cm^{-1}$). In these units the infrared range is 1 to 100 μm, or 10,000 to 100 $cm^{-1}$.

We know that a harmonic oscillator has a vibration frequency equal to

$$v_0 = \frac{1}{2\pi}\left(\frac{k}{\mu}\right)^{1/2}$$

where *k* is the force constant for the oscillator and μ is its reduced mass. This equation applies approximately to vibrations in molecules. It states that stronger bonds with larger force constants have higher vibrational frequencies, and that if the vibration involves a proton attached to a heavy atom, the frequency will decrease by a factor of $\sqrt{2}$ on substitution of hydrogen (mass $m_H$) by deuterium ($m_D$).

$$\frac{v_0(H)}{v_0(D)} = \sqrt{\frac{m_D}{m_H}} = \sqrt{2}$$

This is useful in assigning spectra to vibrations in a molecule.

There are two principal ways to measure vibrational spectra: infrared absorption and Raman scattering. (see Roman Spectroscopy)

Fig. 1 illustrates at the left the transitions directly measured in infrared absorption. There are selection rules that describe the allowed transitions. A vibration must cause a change of electric dipole moment of a molecule for there to be absorption of light. Thus $N_2$ and $O_2$ do not absorb in the infrared, but $H_2O$ and $CO_2$ do. $N_2$ and $O_2$ have no electric dipole moment and vibration leaves the dipole zero. $H_2O$ does have an electric pole moment and vibrations change its magnitude. $CO_2$ has no electric dipole moment, but an asymmetric stretch of its C=O bonds, or an O=C=O bend will break its symmetry and lead to infrared absorption. The infrared spectrum is measured in a spectrophotometer just like those used in

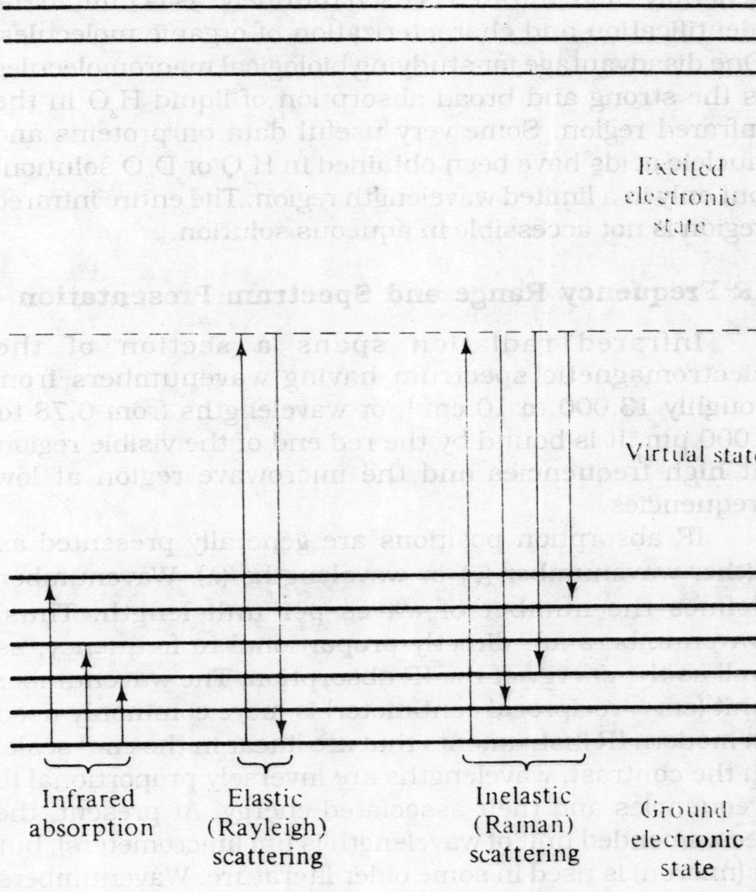


**Fig. 1. Energy-level diagram for infrared absorption, elastic (Rayleigh) scattering, and inelastic (Raman) scattering. Transitions between vibrational energy levels of the ground electronic state absorb infrared radiation; not all transitions are allowed. For elastic scattering the frequency of the incident light is equal to the frequency of the scattered light. For inelastic scattering the frequency of the scattered light is different from that of the incident light. The virtual state shown can have any energy; the scattering of light occurs in any region of the spectrum.**

the visible and ultraviolet region. The Beer-Lambert law is generally valid. Infrared is used routinely, as is nmr, in the identification and characterization of organic molecules. One disadvantage for studying biological macromolecules is the strong and broad absorption of liquid $H_2O$ in the infrared region. Some very useful data on proteins and nucleic acids have been obtained in $H_2O$ or $D_2O$ solution, but only in a limited wavelength region. The entire infrared region is not accessible in aqueous solution.

## IR Frequency Range and Spectrum Presentation

Infrared radiation spans a section of the electromagnetic spectrum having wavenumbers from roughly 13,000 to 10 $cm^{-1}$, or wavelengths from 0.78 to 1000 μm. It is bound by the red end of the visible region at high frequencies and the microwave region at low frequencies.

IR absorption positions are generally presented as either wavenumber ($\bar{v}$) or wavelengths (λ). Wavenumber defines the number of waves per unit length. Thus, wavenumbers are directly proportional to frequency, as well as the energy of the IR absorption. The wavenumber unit ($cm^{-1}$, reciprocal centimeter) is more commonly used in modern IR instruments that are linear in the $cm^{-1}$ scale. In the contrast, wavelengths are inversely proportional to frequencies and their associated energy. At present, the recommended unit of wavelength is μm (micrometers), but μ (micron) is used in some older literature. Wavenumbers and wavelengths can be interconverted using the following equation:

$$\bar{v}\,(\text{in cm}^{-1}) = \frac{1}{\lambda(\text{in mm})} \times 10^4 \qquad (1)$$

IR absorption information is generally presented in the form of a spectrum with wavelength or wavenumber as the x-axis and absorption intensity or percent transmittance as the y-axis (Fig. 2). Transmittance, *T*, is the ratio of radiant power transmitted by the sample (*I*) to the radiant power incident on the sample ($I_0$). Absorbance ($A$) is the logarithm to the base 10 of the reciprocal of the transmittance ($T$).

$$A = \log_{10}(1/T) = -\log_{10} T = -\log_{10} I/I_0 \quad (2)$$

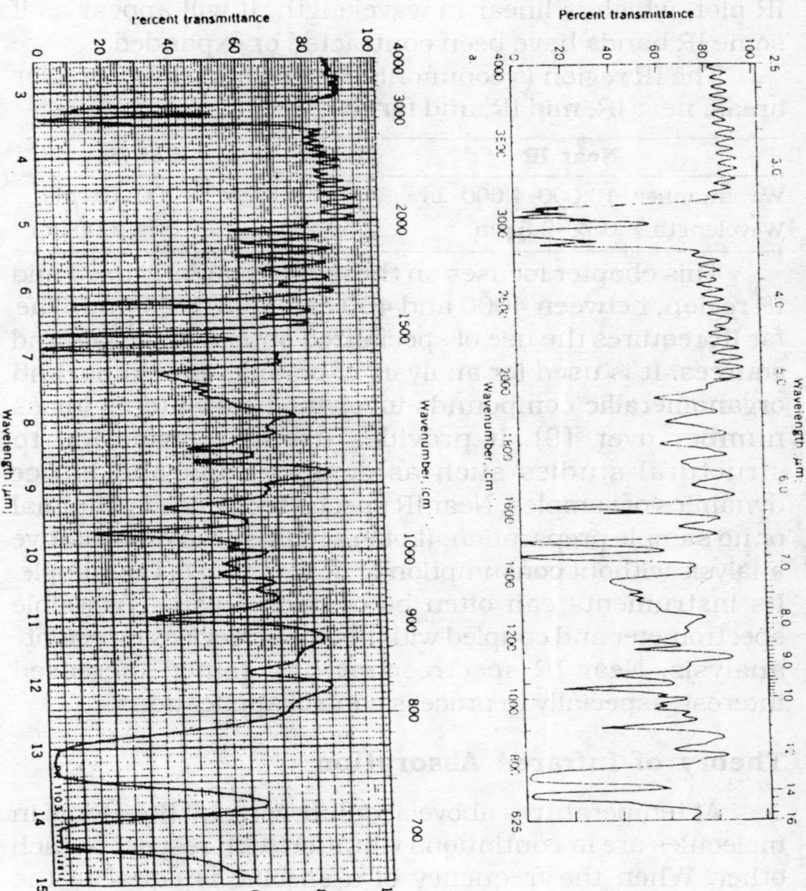

**Fig. 2. IR spectra of polystyrene film with different x-axis units. (a) Linear in wavenumber ($cm^{-1}$), (b) linear in wavelength (μm).**

The transmittance spectra provide better contrast between intensities of strong and weak bands because transmittance ranges from 0 to 100% *T* whereas absorbance ranges from infinity to zero. The analyst should be aware that the same sample will give quite different profiles for the IR spectrum, which is linear in wavenumber, and the IR plot, which is linear in wavelength. It will appear as if some IR bands have been contracted or expanded.

The IR region is commonly divided into three smaller areas: near IR, mid IR, and far IR.

| | **Near IR** | **Mid IR** | **Far IR** |
|---|---|---|---|
| Wavenumber | 13,000–4,000 $cm^{-1}$ | 4,000–200 $cm^{-1}$ | 200–10 $cm^{-1}$ |
| Wavelength | 0.78–2.5 μm | 2.5–50 μm | 50–1,000 μm |

This chapter focuses on the most frequently used mid IR region, between 4000 and 400 $cm^{-1}$ (2.5 to 25 μm). The far IR requires the use of specialized optical materials and sources. It is used for analysis of organic, inorganic, and organometallic compounds involving heavy atoms (mass number over 19). It provides useful information to structural studies such as conformation and lattice dynamics of samples. Near IR spectroscopy needs minimal or no sample preparation. It offers high-speed quantitative analysis without consumption or destruction of the sample. Its instruments can often be combined with UV-visible spectrometer and coupled with fiberoptic devices for remote analysis. Near IR spectroscopy has gained increased interest, especially in process control applications.

## Theory of Infrared Absorption

At temperatures above absolute zero, all the atoms in molecules are in continuous vibration with respect to each other. When the frequency of a specific is equal to the frequency of the IR radiation directed on the molecule the molecule absorbs the radiation.

Each atom has three degrees of freedom corresponding to motions along any of the three Cartesian coordinate axes (x, y, z). A polyatomic molecule of *n* atoms has 3*n* total degrees of freedom. However, 3 degrees of freedom are required to describe translation, the motion of the entire molecule through space. Additionally, 3 degrees of freedom correspond to rotation of the entire molecule. Therefore, the remaining $3n$–6 degrees of freedom are true, fundamental vibrations for nonlinear molecules. Linear molecules possess $3n$–5 fundamental vibrational modes because only 2 degrees of freedom are sufficient to describe rotation. Among the $3n$–6 or $3n$–5 fundamental vibrations (also known as normal modes of vibration) those that produce a net change in the dipole moment may result in an IR activity and those that give polarizability changes may give rise to Raman activity. Naturally some vibrations can be both IR– and Raman active.

The total number of observed absorption bands is generally different from the total number of fundamental vibrations. It is reduced because some modes are not IR active and a single frequency can cause more than one mode of motion to occur. Conversely additional bands are generated by the appearance of overtones (integral multiples of the fundamental absorption frequencies) combinations of fundamental frequencies, differences of fundamental frequencies, coupling interactions of two fundamental absorption frequencies and coupling interactions between fundamental vibrations and overtones or combination bands (Fermi resonance). The intensities of overtone, combination, and difference bands are less than those of the fundamental bands. The combination and blending of all the factors thus create a unique IR spectrum for each compound.

The major types of molecular vibrations are stretching and bending. The various types of vibrations are illustrated in Fig. 3. Infrared radiation is absorbed and the associated energy is converted into these type of motions. The absorption involves discrete, quantized energy levels. However, the individual vibrational motion is usually accompanied by other rotational motions. These combinations lead to the absorption bands, not the discrete line, commonly observed in the mid IR region.

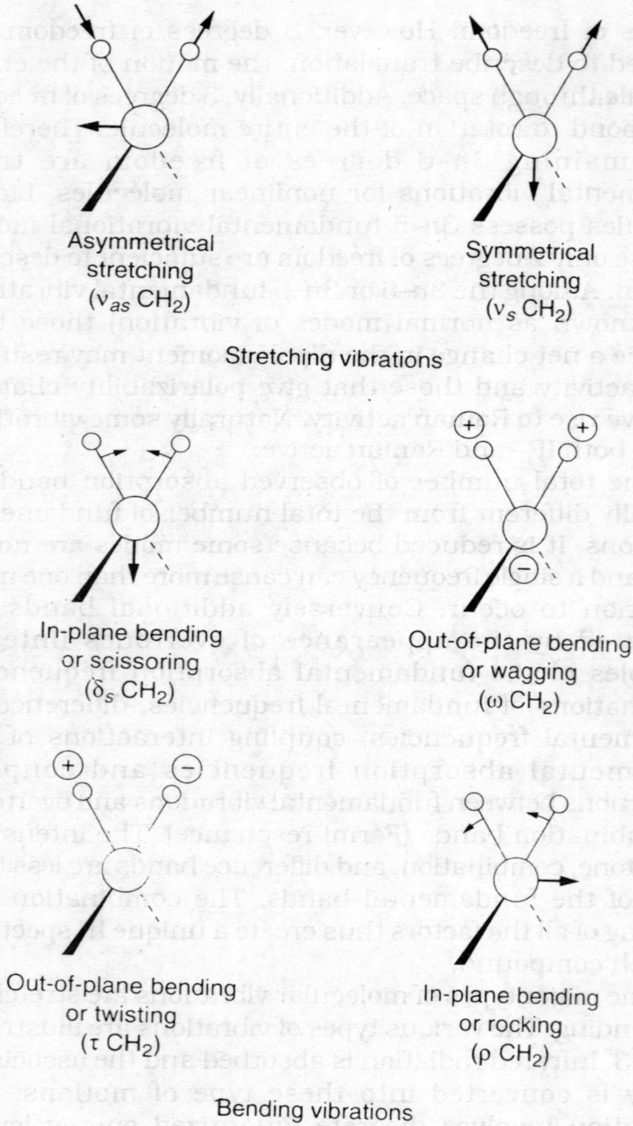


**Fig. 3. Major vibrational modes for a nonlinear group CH$_2$. (+ indicates motion from the plane of page toward reader; − indicates motion from the plane of page away reader).**

Absorption of radiation in the infrared region of the electromagnetic spectrum results in changes in the vibration energy of molecules. Energy changes are typically $6 \times 10^3$ to $42 \times 10^3$ J $mol^{-1}$, which corresponds to 250 to 4,000 $cm^{-1}$ (2.5 to 40 μm), although some occur between 4,000 $cm^{-1}$ and the beginning of the visible region at about 12,500 $Cm^{-1}$ (0.8 μm or 800 nm) in a region known as the near infrared (NIR). A molecule can absorb energy only if there is a net change in the *dipole moment* during a particular vibration, a condition fulfilled by virtually all polyatomic molecules. The absorption spectra of such molecules are often very complex and the underlying reason for this complexity is best understood by first considering the spectrum of a heteronuclear diatomic molecule in the gaseous state.

## Diatomic Molecules

The classical analogy of a vibrating diatomic molecule is that of two weights connected by a spring. The potential energy of such a vibrating system is related to the displacement of the weights relative to each other along the axis of the spring. If the motion is *simple harmonic* the relation is given by $E = \frac{1}{2} fx^2$ where *E* denotes potential energy, *x* the displacement and *f* the force constant or stiffness of the spring. The equation describes a parabola, fig. 4, and shows that identical changes in potential energy occur on stretching or compressing the spring. A diatomic molecule behaves rather differently in that the forces of internuclear repulsion arising when the bond is compressed build up rapidly whilst, on stretching, the bond weakens and may finally disrupt. The resulting vibrational motion is *an harmonic* in nature and the mathematical relation between potential energy and displacement is necessarily more complex. The corresponding curve is modified in the manner shown in figure 4.

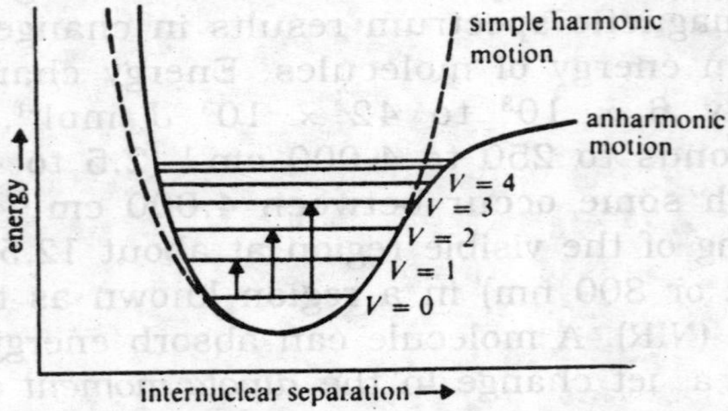


**Fig. 4 : Vibrational energy levels of a diatomic molecule**

The concept of discrete or quantized energy levels can be superimposed on this diagram by representing them as a series of horizontal lines, the spacing of which becomes closer with increasing energy due to the an-harmonic nature of the vibration. In quantum mechanical terms, these levels are labelled $V = 0, 1, 2, 3...$, where $V$ is the vibrational quantum number, and the general mathematical expression for the potential energy of the system expressed in wavenumbers is given by

$$\varepsilon_{vib} = \left(V + \tfrac{1}{2}\right)\sigma_e - \left(V + \tfrac{1}{2}\right)^2 \sigma_e \chi_e \text{cm}^{-1} \qquad (1)$$

$\sigma_e$ denoting the equilibrium vibration frequency and $\chi_e$ the anharmonicity constant. This expression may be used to predict the appearance of the absorption spectrum of a diatomic molecule. At room temperature most molecules are in the ground vibrational state ($V = 0$) so that only transitions from this level need be considered. The spectroscopic selection rule allows any change in the value of $V$, so for transitions originating in the ground state,

$$\Delta\varepsilon_{0\to n} = n\sigma_e[1 - (n+1)\chi_e]\text{cm}^{-1} \qquad (2)$$

where $n$ may be 1,2,3, ... .

Substitution of appropriate values of $n$ into the equation leads to a set of expressions giving the wavenumber of each transition, e.g., the first three and their designations are

$$\sigma_{spec} = \Delta\varepsilon_{0\to1} = \sigma_e(1-2\chi_e) \textit{ fundamental}$$

$$\sigma_{spec} = \Delta\varepsilon_{0\to2} = 2\sigma_e(1-3\chi_e) \textit{ first overtone}$$

$$\sigma_{spec} = \Delta\varepsilon_{0\to3} = 3\sigma_e(1-4\chi_e) \textit{ second overtone}$$

where $\sigma_{spec}$ is the wavenumber of the spectroscopic absorption.

The transition probability falls off rapidly with increasing *n*, and the spectrum therefore consists of two or three lines of diminishing intensity, each one being separated from the next by rather less than the value of $\sigma_e$. Although basically a simple spectrum simultaneous rotational transitions produce a pattern of fine structure lines on either side of the vibrational transition. Rotational fine structure is not usually resolved for sample run in the liquid state or in solution. This is due to collisional broadening and the resulting spectrum would have the appearance shown in fig. 5. For a given molecule, and using the classical analogy considered earlier, the positions of the fundamental and hence the overtones, are determined by the atomic masses $m_1$ and $m_2$, and the stiffness of the bond. These are related by the equation

$$\sigma_e = \frac{1}{2\pi c}\left(\frac{f}{\mu}\right)^{\frac{1}{2}} \qquad (3)$$

where *f* is the force constant of the bond and μ the reduced mass, given by $(m_1 m_2)/(m_1+m_2)$.

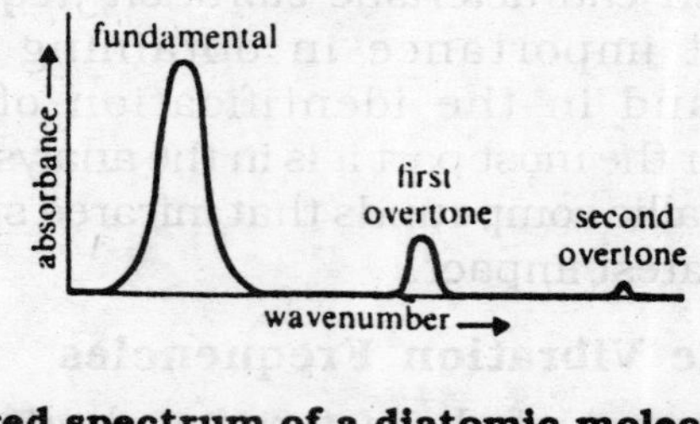


**Fig. 5. Infrared spectrum of a diatomic molecule in solution.**

It is important to note that the stiffer the bond or the lighter the atoms, the higher is the wavenumber of the fundamental and vice versa. Qualitative conclusions can therefore be reached from the relative positions of absorption bands in the infrared region (*vide infra*).

## Polyatomic Molecules

For polyatomic molecules, whose spectra are usually recorded as liquids, solids or in solution, the complexity of the spectrum increases rapidly with the number of atoms $N$ in the molecule. The number of fundamentals, or *normal modes* of vibration is given by $3N$–6 and $3N$–5 for bent and linear molecules respectively. A *normal mode* of vibration is defined as the movement of all the atoms of the molecule in phase. In the case of polyatomics, the number of fundamentals includes both *stretching* and *bending* vibrations, the latter involving changes of bond angle. There are associated overtones with each fundamental and, in addition, other absorption bands may arise due to interactions between certain of the fundamentals and overtones. These are known as *combination* and *Ferni resonance* bands. The overall appearance of the spectrum is that of a large number of resolved and partially resolved bands of varying intensity, figs. 22 to 29. The most intense bands are fundamentals in which there is a large change of dipole moment during the vibration. Conversely, a band may be weak or absent if the dipole change is small or zero. Many bands can be assigned to the vibration of particular chemical groups within a molecule, the wavenumber being largely independent of the rest of the structure. Such *characteristic vibration frequencies* are of paramount importance in obtaining structural information and in the identification of unknown compounds. For the most part it is in the analysis of organic and organometallic compounds that infrared spectrometry has had its greatest impact.

## Characteristic Vibration Frequencies

The occurrence of characteristic vibration or group frequencies can be explained in terms of relative masses and of force constants using the classical analogy of weights and springs as depicted in fig. 6.

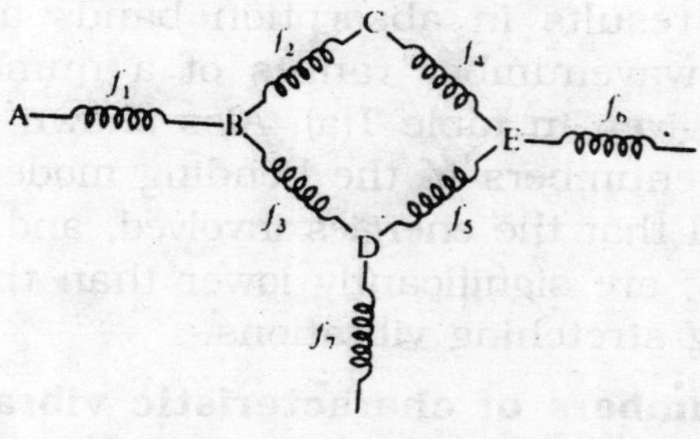


**Fig. 6. Classical analogy of polyatomic molecule.**

**(a) Mass Effect**

It the masses *A, B, C, D* ... are all similar and the force constants $f_1$, $f_2$, $f_3$... are of the same magnitude, the vibrations of individual atoms are strongly coupled with the result that no band can be assigned solely to any particular group of atoms. If, however, the mass of atom A is considerably smaller than those of the other atoms, one mode of vibration will involve the stretching of the bond between A and the rest of the molecule. The system can be considered as approximating to a diatomic molecule A—X, and the wavenumber of the vibration is then given by

$$\sigma_{A-X} = \frac{1}{2\pi c}\left(\frac{f_{A-X}}{\mu_{A-X}}\right)^{\frac{1}{2}}$$

where $f_{A\text{-}X}$ is the force constant of the A—X bond and

$$\mu_{A-X} = \left(\frac{AX}{A+X}\right) \simeq A$$

Hence, to a good approximation

$$\sigma_{A-X} = \frac{1}{2\pi c}\left(\frac{f_{A-X}}{A}\right)^{\frac{1}{2}}$$

and the wavenumber of the vibration is therefore independent of the masses of the other atoms in the molecule. This type of stretching vibration is examplified by hydrogen atoms linked to carbon, oxygen, nitrogen or sulphur, and results in absorption bands around 3,000 $cm^{-1}$. The wavenumber ranges of a number of such groups are given in table 1(a). Also shown in the table are the wavenumbers of the bending modes, and it should be noted that the energies involved, and hence the wavenumbers, are significantly lower than those of the corresponding stretching vibrations.

**Table 1. Wavenumbers of characteristic vibrations**

**(a) X—H Vibrations**

| X—H | Stretch/$cm^{-1}$ | Bend/$cm^{-1}$ |
|---|---|---|
| C—H | 2700–3300 | 1300–1500<br>600–900 |
| O—H | 3000–3700 | 1200–1500 |
| N—H | 3000–3700 | 1500–1700 |
| S—H | 2550–2600 | 700–900 |

**(b) Saturated and unsaturated groups**

| Group | Stretch/$cm^{-1}$ |
|---|---|
| C—C, C—N, C—O | 400–1300 |
| C=C, C=N, C=O | 1550–1900 |
| C ≡ C, C ≡ N | 2100–2400 |

**(b) Force Constant Effect**

Considering the same model as that shown in fig. 6, if the force constant of the A—B bond is significantly higher than those for the rest of the molecule, one stretching vibration will occur at a higher wavenumber as $\sigma_e$ is directly proportional to $f_{A-B}$. This situation arises in molecules with unsaturated groups such as C=C,C ≡ C and C=O because the magnitude of the force constant increases significantly with bond order. Examples of the wavenumber ranges for some unsaturated and saturated groups are compared in table 1(b). Differences between the force constants of single bonds are not sufficiently great to enable characteristic group frequencies to be identified.

Prominent bands can be observed, however, for groups such as C—O where the dipole change during the vibration is large. For groups with determined by the position of the group in the molecule. Thus little or no absorption due to the multiple bond is observed for symmetrically substituted alkenes or alkynes, whereas a terminal unsaturated group gives rise to a strong absorption.

## Factor Affecting Group Frequencies

### Hydrogen Bonding

Compounds containing proton donor groups such as O—H and N—H can be involved in intra- or intermolecular hydrogen bonding in the presence of proton acceptors, e.g., O, N, halogens, C=C. The stiffness of the X—H bond is thereby lessened resulting in a lowering of the stretching frequency, and the band broadens and often intensifies. Conversely, the frequency of the bending mode is raised but the effect is much less pronounced. *Intermolecular* hydrogen bonding is suppressed at elevated temperatures but it is favoured by a high solute concentration. *Intramolecular* hydrogen bonding is also reduced at elevated temperatures but it is unaffected by changes in solute concentration. These effects are particularly significant in the spectra of alcohols, phenols, carboxylic acid and amines.

### Adjacent groups

The wealth of structural information that can be obtained from infrared spectra arises largely from differences in band position, shape and intensity that are observed when comparing similar but not identical compounds. Predictable shifts occur on altering the substituents or structure adjacent to a group giving rise to a particular absorption. These can be explained in terms of an *inductive effect*, a *mesomeric* or *resonance effect* and *steric factors* such as ring strain. The effects are typified by the carbonyl group, whose stretching frequency occurs within the range 1650 to 1850 $cm^{-1}$. Shift are usually related to the position of the carbonyl band in pure acetone (1715 $cm^{-1}$). The proximity of electron withdrawing substituents causes an inductive withdrawal of electron density from around the oxygen atom thereby shortening the bond, fig. 7. This increases the force constant and hence the wave number of the vibration, table 2. If the adjacent group contains a weakly electronegative atom with lone-pair electrons or electrons occupying π-orbitals, resonance interaction with electrons in the carbonyl group leads to a reduction in bond, fig. 7. The corresponding smaller force constant results in a lower value for the wavenumber of the vibration, table 2. Steric hindrance may inhibit the resonance effect by reducing the coplanarity of a conjugated system whilst carbonyl groups which form part of a strained ring system absorb at high wavenumbers because of the constraint imposed upon the carbon atom, table 2.

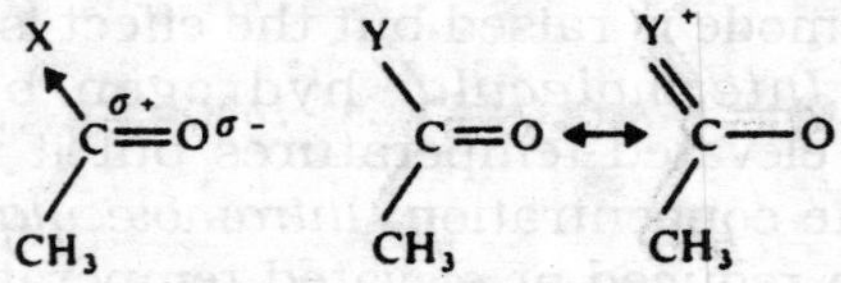

**Fig. 7. Effect of substituents of the carbonyl bond. (a) Inductive effect, X=H, OH, OR, halogen. (b) Resonance effect, Y = $NH_2$, NHR, $NR_2$, phenyl.**

**Table 2. Effect of substituents on the wave number of the carbonyl stretching vibration.**

| Compound | C=O Strectch/$cm^{-1}$ |
|---|---|
| saturated ketones | 1710–1720 |
| carboxylic acids, monomer | 1750–1770 |
| carboxylic acids, dimer | 1710–1720 |
| saturated esters | 1725–1745 |
| acid chlorides | 1790–1815 |
| aliphatic amides | 1630–1700 |
| aromatic ketones | 1680–1700 |
| aromatic and α,β-unsaturated esters | 1715–1730 |
| γ-lactones | 1760–1780 |

## Coupled Vibrations

Groups in which two or more atoms of an element are bonded to a common atom are mechanically coupled resulting in a band which is generally resolved into two components. These correspond to *symmetric* and *asymmetric* modes of vibration. They also execute a variety of bending vibrations each of which may be observed at a characteristic wavenumber. Examples include $CH_3$, $CH_2$, $NH_2$, $NO_2$ and COO–. The stretching and bending modes of a methylene group are illustrated in fig. 8. It should be noted that asymmetric vibrations occur at a higher wavenumber than symmetric and that the absorption is more intense. Another form of vibrational coupling occurs between the C—H groups in both aromatic and olefinic compounds resulting in bands which are characteristic of the substitution pattern.; In the case of *out-of-plane bending,* they are of a high intensity and are especially useful for diagnostic purposes (*vide infra*).

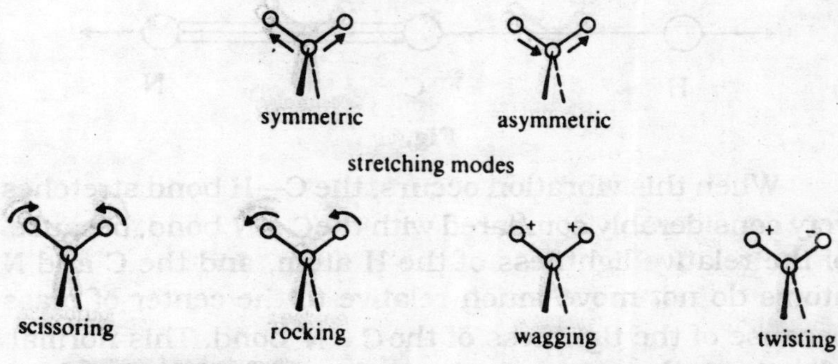


**Fig. 8. Stretching and bending modes of a methylene group.**

## Characteristic Group Frequencies

The complete analysis of the infrared spectrum of a molecule containing more than a dozen atoms is very difficult, and it is sometimes impossible. However, very useful information can be obtained from the spectra of quite large molecules if one makes use of simple theory and empirical correlations. In particular, individual functional groups in molecules absorb in characteristic regions of the infrared spectrum. For example, a molecule containing the alcohol group, O—H, always shows an absorption in the 3100–3500 cm$^{-1}$ region of the spectrum, while molecules containing the carbonyl group C=O characteristically absorb in the 1600–1800 cm$^{-1}$ region. Table 3 gives a few wavenumbers and wavelengths for different kinds of bonds.

There are two main reasons why these convenient simplifications occur. If a bond involves a hydrogen atom attached to a much heavier atom such as an oxygen atom, some of the normal modes of vibration involve the very light atom moving with respect to the rest of the molecule, which does not move very much in relation to the center of gravity. A very simple example is provided by the spectrum of HCN. One of the normal-mode wave number is 2089 cm$^{-1}$, and this relates to the symmetric normal mode of vibration:

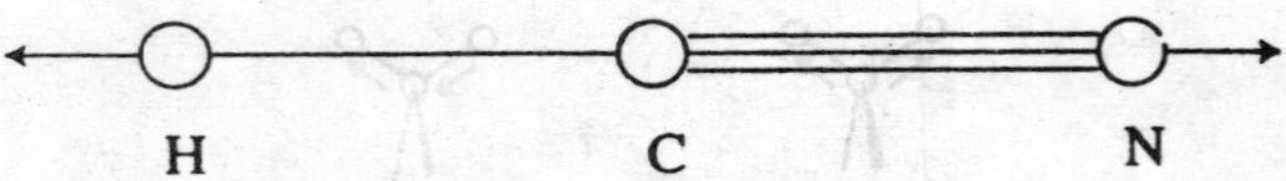


**Fig.**

When this vibration occurs, the C—H bond stretches very considerably compared with the C ≡ N bond, because of the relative lightness of the H atom, and the C and N atoms do not move much relative to the center of mass because of the tightness of the C ≡ N bond. This normal vibration is therefore very close to being simply a stretching of the C—H bond. Similarly, with an alcohol, some of the normal modes of vibration are very close to being the simple stretching of the O—H bond.

Another factor that tends to favor characteristic frequencies is the presence of multiple bonds. Here the effect is due to the wide disparity between the force constant of the multiple bond and the force constants for other bonds in the molecule. A simple example is again provided by

HCN, which has a normal-mode wave number of 3312 $cm^{-1}$ for the antisymmetric vibration:

It happens that the force constant for the $C \equiv N$ bond is over three times that for the H—C bond, and the normal-mode analysis shows that this antisymmetric vibration is very largely the stiff vibration of the $C \equiv N$ bond.

Bonds involving hydrogen atoms and multiple bonds therefore tend to have characteristic group frequencies. There are also some effects that work against characteristic group frequencies; the most important is when there is considerable symmetry so that the molecule contains two or more equivalent bonds. A simple example is provided by the water molecule, which has normal mode frequencies of 3756, 3652, and 1559 $cm^{-1}$. The lowest of these may be attributed to the bending vibration. However, neither of the two higher frequencies can be ascribed to the stretching of one bond, since the two O—H bonds are equivalent. The stretching motions of the two bonds will therefore the strongly coupled, and in these two normal modes the two bonds stretch either in phase or out of phase with one another.

These limitations, however, are more prominent with fairly small molecules. With molecules of any complexity the characteristic group frequencies can provide useful supporting evidence for the presence of specific functional groups. Moreover, because of the large number of vibrational modes in a molecule of any size, the infrared spectrum is very complicated and is unique to the molecule. Fig. 9 shows a typical infrared spectrum; since this was obtained with the substance in the liquid phase, the rotational fine structure is blurred, but the features of this spectrum will to be found with any other molecule and will allow a very convincing identification of the substance to be made. Because the infrared spectrum of a substance is so uniquely characteristic, it is often referred to as a *fingerprint spectrum.*

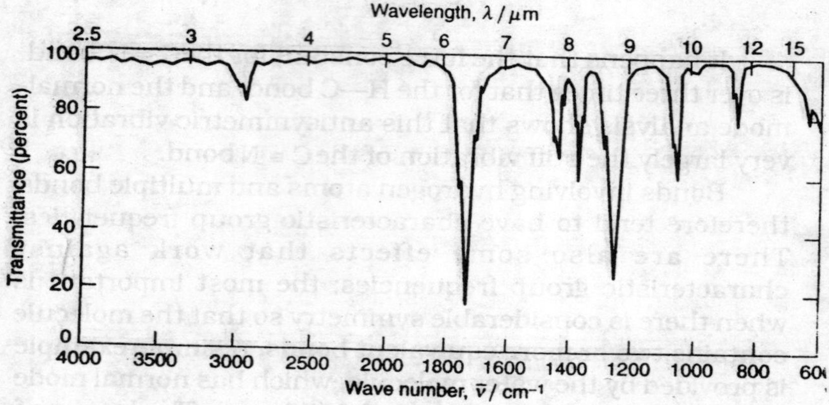


**Fig. 9. The infrared spectrum of methyl acetate. This spectrum was obtained with material in the liquid phase.**

## Symmetry Species Corresponding To Infrared and Raman Spectra

| Symmetry Species | Infrared Spectra | | | Raman Spectra | | | | | |
|---|---|---|---|---|---|---|---|---|---|
| | x | y | z | $x^2$ | $y^2$ | $z^2$ | xy | xz | yz |
| $C_{2v}$ | $b_1$ | $b_2$ | $a_1$ | $a_1$ | $a_1$ | $a_1$ | $a_2$ | $b_1$ | $b_2$ |
| $C_{3v}$ | $e$ | $e$ | $a_1$ | $a_1$, $e$ | $a_1$, $e$ | $a_1$ | $e$ | $e$ | $e$ |
| $C_{\infty v}$ | $\pi$ | $\pi$ | $\sigma^+$ | $\sigma$,$\delta$ | $\sigma^+$,$\delta$ | $\sigma^+$ | $\delta$ | $\pi$ | $\pi$ |
| $D_{\infty h}$ | $\pi_u$ | $\pi_u$ | $\sigma_u^+$ | $\sigma_g^+$, $\delta$ | $\sigma_g^+$, $\delta_g$ | $\sigma_g$ | $\delta_g$ | $\pi_g$ | $\pi_g$ |
| $D_{3h}$ | $e'$ | $e'$ | $a_2''$ | $a'_1$, $e'$ | $a'_1$, $e'$ | $a'_1$ | $e'$ | $e''$ | $e''$ |
| $C_{2h}$ | $b_u$ | $b_u$ | $a_u$ | $a_g$ | $a_g$ | $a_g$ | $a_g$ | $b_g$ | $b_g$ |
| $D_{6h}$ | $e_{1u}$ | $e_{1u}$ | $a_{2u}$ | $a_{1g}$, $e_{2g}$ | $a_{1g}$, $e_{2g}$ | $a_{1g}$ | $e_{2g}$ | $e_{1g}$ | $e_{1g}$ |

**Determination of the vibration-rotation spectrum of hydrogen chloride, and calculation of the force constant, moment of inertia and bond length for HCl.**

*Method.* Place a thin layer of sulphuric acid in a 4 cm quartz or fused silica cell such as is used in ultraviolet spectrophotometry. Add carefully a few crystals of sodium chloride and cover the cell immediately with a polythene cap. Record the absorption curve over the region 3500–2500 $cm^{-1}$ by means of a grating instrument using an empty quartz cell in the reference beam, and

(a) determine the position (in $cm^{-1}$) of the Q branch, i.e., the 'gap' in the spectrum (Fig. 10) and hence calculate $k$ the force constant.

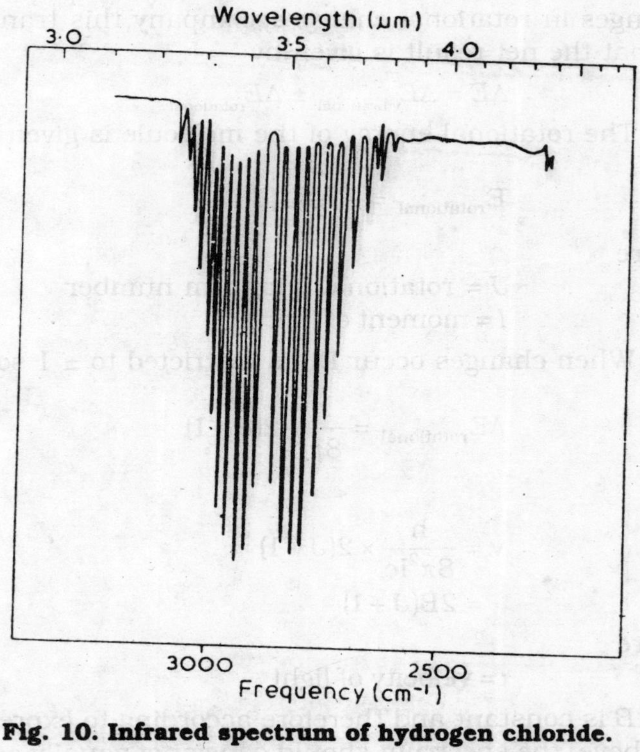


**Fig. 10. Infrared spectrum of hydrogen chloride.**

(b) count the number of bands in the P and R branches and the range in $cm^{-1}$. Calculate the average separation in wave numbers and hence obtain *I*, the moment of inertia.

(c) calculate the bond distance from the relationship $I = \mu r^2$.

Adequate resolution for the complete experiment is obtained with a simple grating instrument or a more expensive large prism machine. If a simple prism instrument is available, the Q branch may still be observed and part (a) of the calculation can be done. The 4 cm quartz cell enables a gas cell to be dispensed with although the latter is essential if HI or HBr are to be examined.

The change in vibrational energy observed in this experiment is due to the transition $V_0 \rightarrow V_1$ (Fig. 10). Changes in rotational energy accompany this transition so that the net result is given by

$$\Delta E = \Delta E_{\text{vibrational}} \pm \Delta E_{\text{rotational}} \qquad (1)$$

The rotational energy of the molecule is given by

$$E_{\text{rotational}} = \frac{h^2}{8\pi^2 I} J(J+1)$$

where

$J$ = rotational quantum number
$I$ = moment of inertia

When changes occur DJ is restricted to ± 1 so that

$$\Delta E_{\text{rotational}} = \frac{h^2}{8\pi^2 I} 2(J+1)$$

or

$$v = \frac{h}{8\pi^2 Ic} \times 2(J+1)$$
$$= 2B(J+1)$$

where

$c$= velocity of light

*B* is constant and therefore according to expression (1) above, the spectrum should consist of equally spaced peaks on each side of a central position corresponding to $\Delta E_{vibrational}$. As a result of the selection rule $\Delta J = \pm 1$, $(J+1)$ may be ±1, ±2, ±3... and therefore absorption corresponding to $\Delta E_{vibrational}$ alone will not be observed. This 'gap' in the spectrum is known as the Q branch and shows clearly in the absorption spectrum obtained in an actual experiment and recorded in Fig. 10. When $(J+1)$ is equal to +1, +2, +3... the R branch is observed, and when it is equal to −1, −2, −3... the P branch is recorded.

The frequency of the radiation associated with the transition $V_0 \rightarrow V_1$ is loosely referred to as the vibrational frequency of the molecule. By treating the two atoms H and Cl as two masses connected by a spring and applying Hooke's law the frequency is shown to be, as a first approximation

$$\bar{v} = \frac{1}{2\pi c}\sqrt{\left(\frac{k}{\mu}\right)}$$

where

$c$ = velocity of light

$k$ = force constant

$\mu$ = reduced mass of system

$$= \frac{m_1 m_2}{m_1 + m_2} \times \frac{1}{N}$$

where

$m_1, m_2$ = respective *atomic weights of the atoms*

$N$ = Avogadro number.

$\bar{v}$ is obtained from the position of the Q branch, hence $k$ may be determined. From the average separation of the peaks, say $x$ cm$^{-1}$

$$x = \frac{h}{8\pi^2 Ic} \times 2 \quad (J + 1 = 1)$$

$$I = \frac{h}{4\pi^2 xc}$$

Now

$$I = \mu r^2$$

where

r = internuclear distance

$$\therefore \quad r^2 = \frac{I}{\mu}$$

$$r = \sqrt{\left(\frac{I}{\mu}\right)}$$

$$= \sqrt{\left\{(I \times N) / \frac{1.008 \times 35.46}{36.47}\right\}}$$

From Fig. 10, it may be seen that the separation between the peaks in the P and R branches is not strictly uniform and therefore not in accord with the simplified discussion presented above. The discrepancy is due to coupling between the rotation and vibration of the molecule. Also the peaks vary in intensity due to the fact that the population of the various rotational energy levels varies in the lower state (*V* = *0*). For details of other factors which complicate the spectra,.

Application of the formula for the frequency of vibration (in $cm^{-1}$) enables *approximate* figures to be obtained from some of the group frequencies, for example for C=O, CH, N—H, and so on. The value of *k* used in the calculation is $5 \times 10^5$, $10 \times 10^5$ and $15 \times 10^5$ dynes/cm for single, double and triple bonds respectively.

**How It Works**

In simple terms, IR spectra are obtained by detecting changes in transmittance (or absorption) intensity as a function of frequency. Most commercial instruments separate and measure IR radiation using dispersive spectrometers of Fourier transform spectrometers.

## Instrumentation

The arrangement of a typical infrared spectrophotomer is shown in diagrammatic form in Fig. (11).

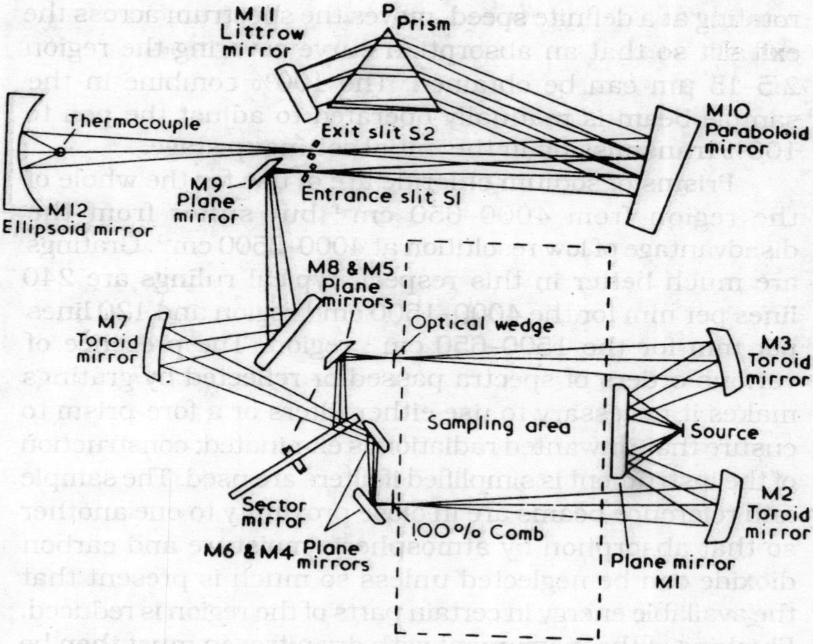


**Fig. 11. Optics of infrared spectrophotometer.**

The various components of a spectrophotometer, viz. light source, monochromator, and detector are easily recognized. The rotating sector mirror (Fig. 11) reflects the reference beam from M3 into the apparatus during one half of its rotation and allows the sample beam from M2 through during the other half. Any difference between the two beams causes an out-of-balance signal; at balance the two beams have the same intensity and produce a DC signal but when unbalanced an AC signal results which causes movement of the optical wedge into or out of the reference beam until the two beams are balanced. The optical wedge is coupled to a pen and an absorption curve is drawn as the spectrum is scanned. This system is known as the optical null method. The wavelength of the radiation emerging from the exit slit depends upon the position of the Littrow mirror M11 relative to the prisms. This mirror, rotating at a definite speed, moves the spectrum across the exit slit so that an absorption curve covering the region 2.5–15 μm can be obtained. The 100% combine in the sample beam is manually operated to adjust the pen to 100% transmission in the initial settingup stage.

Prisms of sodium chloride are of use for the whole of the region from 4000–650 $cm^{-1}$ but suffer from the disadvantage of low resolution at 4000–2500 $cm^{-1}$. Gratings are much better in this respect; typical rulings are 240 lines per mm for the 4000–1500 $cm^{-1}$ region and 120 lines per mm for the 1500–650 $cm^{-1}$ region. The presence of various orders of spectra passed or reflected by gratings makes it necessary to use either filters or a fore-prism to ensure that unwanted radiation is eliminated; construction of the instrument is simplified if filters are used. The sample and reference beams are in close proximity to one another so that absorption by atmospheric moisture and carbon dioxide can be neglected unless so much is present that the available energy in certain parts of the region is reduced. Flushing of the instrument with dry nitrogen must then be carried out.

## Dispersive Spectrometers

Dispersive spectrometers, introduced in the mid-1940s and widely used since, provided the robust instrumentation required for the extensive application of this technique.

## Spectrometer Components

An IR spectrometer consists of three basic components: radiation source, monochromator, and detector. A schematic diagram of a typical dispersive spectrometer is shown in Fig. 12.

The common radiation source for the IR spectrometer is an inert solid heated electrically to 1000 to 1800°C. Three popular types of sources are Nernst glower (constructed or rare-earth oxides), Globar (constructed of silicon carbide), and Nichrome coil. They all produce continuous radiations, but with different radiation energy profiles.

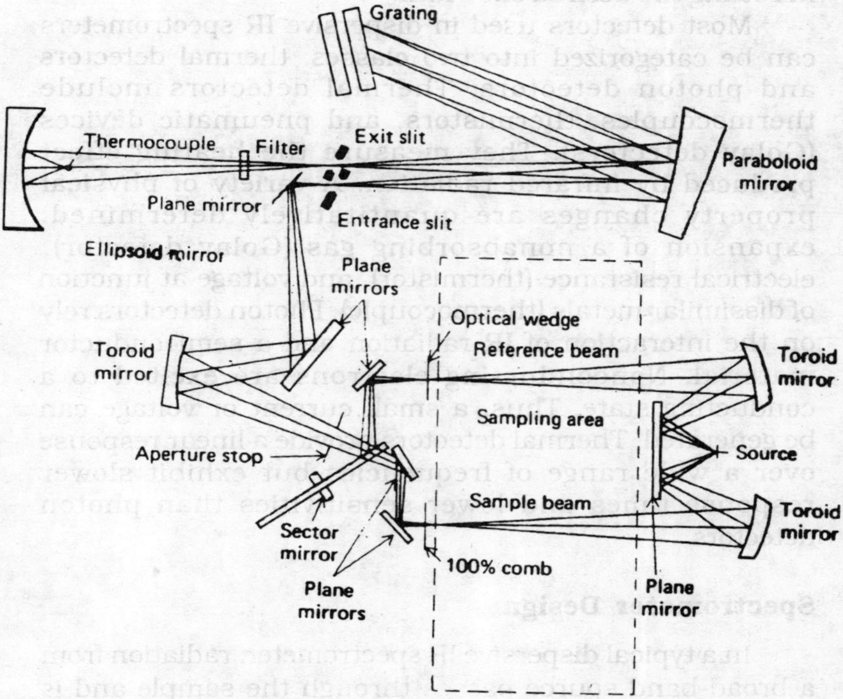


**Fig. 12. Schematic diagram of a commercial dispersive IR instrument, the Perkin-Elmer Model 237B Infrared Spectrometer.**

The monochromator is a device used to disperse a broad spectrum of radiation and provide a continuous calibrated series of electromagnetic energy bands of determinable wavelength or frequency range. Prisms or gratings are the dispersive components used in conjunction with variable slit mechanisms, mirrors and filters. For example, a grating rotates to focus a narrow band of frequencies on a mechanical slit. Narrower slits enable the instrument to better distinguish more closely spaced frequencies of radiation, resulting in better resolution. Wider slits allow more light to reach the detector and provide better system sensitivity. Thus, certain compromise is exercised in setting the desired slit width.

Most detectors used in dispersive IR spectrometers can be categorized into two classes: thermal detectors and photon detectors. Thermal detectors include thermocouples, thermistors, and pneumatic devices (Golay detectors). They measure the heating effect produced by infrared radiation. A variety of physical property changes are quantitatively determined: expansion of a nonabsorbing gas (Golay detector), electrical resistance (thermistor), and voltage at junction of dissimilar metals (thermocouple). Photon detectors rely on the interaction of IR radiation and a semiconductor material. Nonconducting electrons are excited to a conducting state. Thus, a small current or voltage can be generated. Thermal detectors provide a linear response over a wide range of frequencies but exhibit slower response times and lower sensitivities than photon detectors.

## Spectrometer Design

In a typical dispersive IR spectrometer, radiation from a broad-band source passes through the sample and is dispersed by a monochromator into component frequencies (Fig 12). Then the beams fall on the detector, which generates an electrical signal and results in a recorder response.

Most dispersive spectrometers have a double-beam design. Two equivalent beams from the same source pass through the sample and reference chambers respectively. Using an optical chopper (such as a sector mirror), the reference and sample beams are alternately focused on the detector. Commonly, the change of IR radiation intensity due to absorption by the sample is detected as an off-null signal that is translated into the recorder response through the actions synchronous motors.

## Fourier Transform Spectrometers

Fourier transform spectrometers have recently replaced dispersive instruments for most applications due to their superior speed and sensitivity. They have greatly extended the capabilities of infrared spectroscopy and have been applied to many areas the are very difficult or nearly impossible to analyze by dispersive instruments. Instead of viewing each component frequency sequentially, as in a dispersive IR spectrometer, all frequencies are examined simultaneously in Fourier transform infrared (FTIR) spectroscopy.

### Spectrometer Components

There are three basic spectrometer components in an FT system: radiation source, interferometer, and detector. A simplified optical layout of a typical FTIR spectrometer is illustrated in Fig. 13.

The same types of radiation sources are used for both dispersive and Fourier transform spectrometers. However, the source is more often water-cooled in FTIR instruments to provide better power and stability.

In contrast, a completely different approach is taken in an FTIR spectrometer to differentiate and measure the absorption at component frequencies. The monochromator is replaced by an interferometer, which divides radiant beams, generates an optical path difference between the beams, the recombines them in order to produce repetitive interference signals measured as a function of optical path difference by a detector. As its name implies, the interferometer produces interference signals, which contain infrared spectral information generated after passing through a sample.

The most commonly used interferometer is a Michelson interferometer. It consists of three active components: a moving mirror, a fixed mirror, and a beamsplitter (Fig. 13). The two mirrors are perpendicular to each other. The beamsplitter is a semireflecting device and is often made by depositing a thin film of germanium onto a flat KBr substrate. Radiation from the broadband IR source is collimated and directed into the interferometer, and impinges on the beamsplitter. At the beasplitter, half the IR beam is transmitted to the fixed mirror and the remaining half is reflected to the moving mirror. After the divided beams are reflected from the two mirrors, they are recombined at the beamsplitter. Due to changes in the relative position of the moving mirror to the fixed mirror, an interference pattern is generated. The resulting beam then passes through the sample and is eventually focused on the detector.

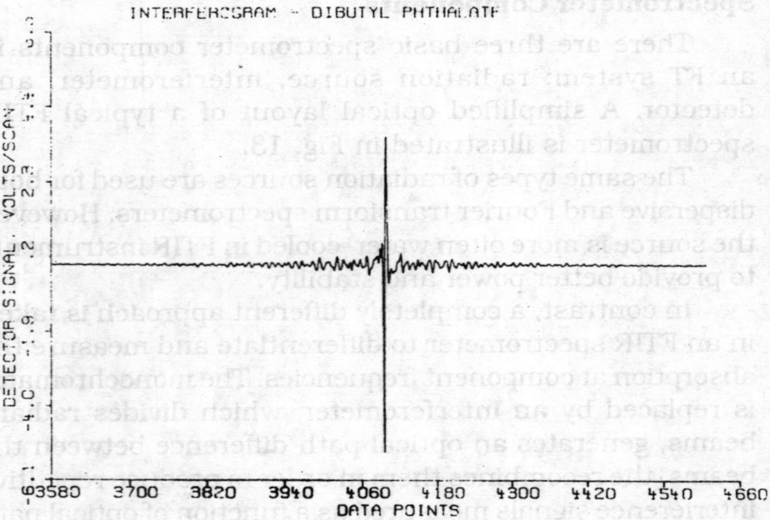


**Fig. 13. Simplified optical layout of typical FTIR spectrometer.**

For an easier explanation, the detector response for a single-frequency component from the IR source is first considered. This simulates an idealized situation where the source is monochromatic, such as a laser source. As previously described, differences in the optical paths between the two split beams are created by varying the relative position of moving mirror to the fixed mirror. If the two arms of the interferometer are of equal length, the two split beams travel through the exact same path length. The two beams are totally in phase with each other; thus, they interfere constructively and lead to a maximum in the detector response. This position of the moving mirror is called the point of zero path difference (ZPD). When the moving mirror travels in either direction by the distance $\lambda/4$, the optical path (beamsplitter-mirror-beamsplitter) is changed by $2(\lambda/4)$, or $\lambda/2$. The two beams are 180° out of phase with each other, and thus interfere destructively. As the moving mirror travels another $\lambda/4$, the optical path difference is now $2(\lambda/2)$, or $\lambda$. The two beams are again in phase with each other and result in another constructive interference.

When the mirror is moved at a constant velocity, the intensity of radiation reaching the detector varies in a sinusoidal manner to produce the interferogram output shown in Fig. 13. The interferogram is the record of the interference signal. It is actually a time domain spectrum and records the detector response changes versus time within the mirror scan. If the sample happens to absorb at this frequency, the amplitude of the sinusoidal wave is reduced by an amount proportional to the amount of sample in the beam.

Extension of the same process to three component frequencies results in a more complex interferogram, which is the summation of three individual modulated waves, as shown in Fig. 14. In contrast to this simple, symmetric interferogram, the interferogram produced with a broadband IR source displays extensive interference patterns. It is a complex summation of wuperimposed sinusodial wavews, each wave corresponding to a single frequency. When this IR beam is directed through the sample, the amplitudes of a set of waves are reduced by absorption if the frequency of this set of waves is the same as one of the characteristic frequencies of the sample (Fig. 15).

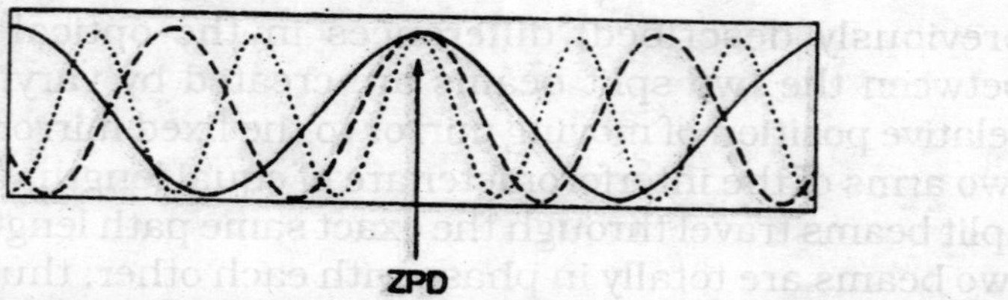


**Fig. 14. Interferogram consisting of three modulated cosine waves. The greatest amplitude occurs at the point of zero path difference (ZPD).**

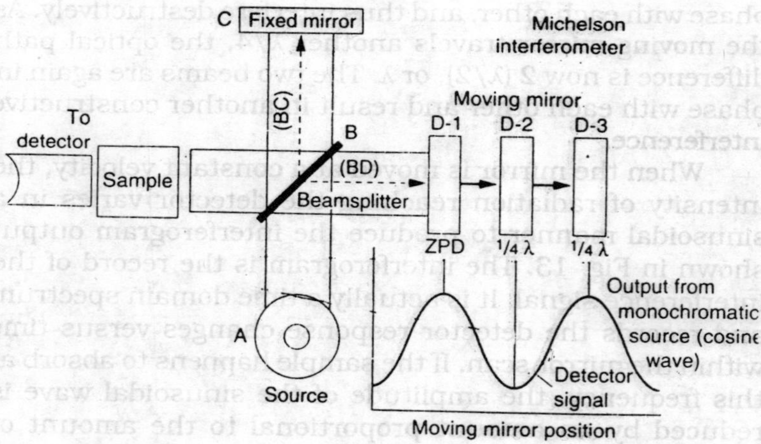


**Fig. 15. A typical interferogram produced with a broadband IR source.**

The interferogram contains information over the entire IR region to which the detector is responsive. A mathematical operation known as Fourier transformation converts the interferogram (a time domain spectrum displaying intensity versus time within the mirror scan) to the final IR spectrum, which is the familiar frequency domain spectrum showing intensity versus frequency. This also explains how the term *Fourier transform infrared* spectrometry is created.

The detector signal is amples at small, precise intervals during the mirror scan. The sampling rate is controlled by an internal, independent reference, a modulated monochromatic beam from a helium neon (HeNe) laser focused on a separate detector.

The two most popular detectors for a FTIR spectrometer are deuterated triglycine sulfate (DTGS) and mercury cadmium telluride (MCT). The response times of many detectors (for example, thermocouple and thermistor) used in dispersive IR instruments are too slow for the rapid scan times (1 sec or less) of the interferometer. The DTGS detector is a pyroelectric detector that delivers rapid responses because it measures the changes in temperature rather than the value of temperature. The MCT detector is a photon (or quantum) detector that depends on the quantum nature of radiation and also exhibits very fast response. Whereas DTGS detectors operate at room temperature, MCT detectors must be maintained at liquid nitrogen temperature (77°K) to be effective. In general, the MCT detector is faster and more sensitive than the DTGS detector.

**Spectrometer Design**

The basic instrument design is quite simple. Fig. 16 illustrates the design of a typical FTIR spectrometer. The IR radiation from a broadband source is first directed into an interferometer, where it is divided and then recombined after the split beams travel different optical paths to generate constructive and destructive interference. Next, the resulting beam passes through the sample compartment and reaches to the detector.

Most benchtop FTIR spectrometers are single-beam instruments. Unlike double-beam grating spectrometers, single-beam FTIR does not obtain transmittance or absorbance IR spectra in real time. A typical operating procedure is described as follows :

1. A background spectrum (Fig. 17) is first obtained by collecting an interferogram (raw data), followed by processing the data by Fourier transform

conversion. This is a response curve of the spectrometer, and takes account of the combined performance of source, interferometer, and detector. The background spectrum also includes the contribution from any ambient water (two irregular groups of lines at about 3600 $cm^{-1}$ and about 1600 $cm^{-1}$) and carbon dioxide (doublet at 2360 $cm^{-1}$ and sharp spike at 667 $cm^{-1}$) present in the optical bench.

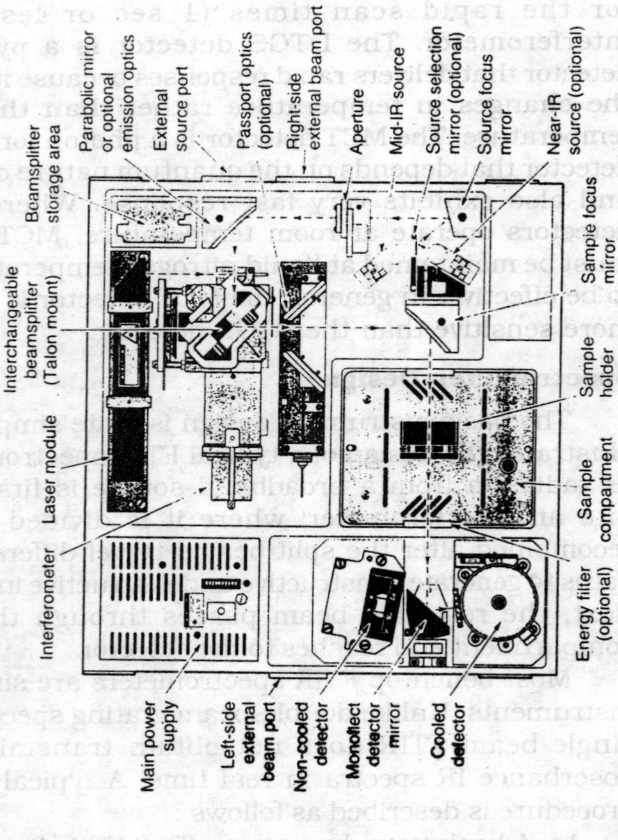


**Fig. 16. Schematic diagram of the Nicolet Magna-IR 750 FTIR Spectrometer. (Reprinted by permission of Nicolet Instrument Corporation).**

2. Next, a single-beam sample spectrum is collected (Fig. 18). It contains absorption bands from the sample and the background (air or solvent).
3. The ratio of the single-beam sample spectrum in Fig. 18 against the single beam back-ground spectrum in Fig. 17 results in a "double-beam" spectrum of the sample (Fig. 19).

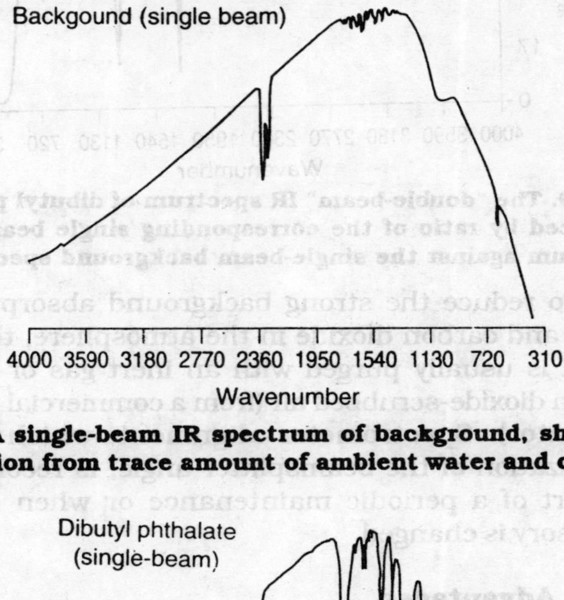


**Fig. 17. A single-beam IR spectrum of background, showing contribution from trace amount of ambient water and carbon dioxide.**

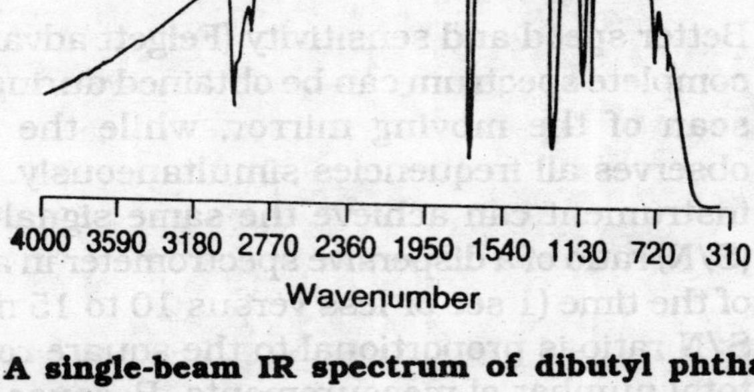


**Fig. 18. A single-beam IR spectrum of dibutyl phthalate (a liquid sample).**

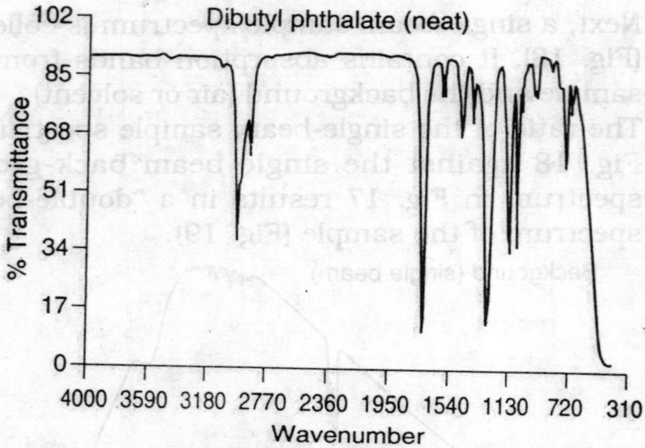


**Fig. 19. The "double-beam" IR spectrum of dibutyl phthalate, produced by ratio of the corresponding single beam sample spectrum against the single-beam background spectrum.**

To reduce the strong background absorption from water and carbon dioxide in the atmosphere, the optical bench is usually purged with an inert gas or with dry, carbon dioxide-scrubbed air (from a commercial purge gas generator). Spectrometer alignment, which includes optimization of the beamsplitter angle, is recommended as part of a periodic maintenance or when a sample accessory is changed.

## FTIR Advantages

FTIR instruments have distinct advantages over dispersive spectrometers :

(i) Better speed and sensitivity (Felgett advantage). A complete spectrum can be obtained during a single scan of the moving mirror, while the detector observes all frequencies simultaneously. An FTIR instrument can achieve the same signal-to-noise (S/N) ratio of a dispersive spectrometer in a fraction of the time (1 sec or less versus 10 to 15 min). The S/N ratio is proportional to the square root of the total number of measurements. Because multiple spectra can be readily collected in 1 min or less, sensitivity can be greatly improved by increasing S/N through coaddition of many repeated scans.

(ii) Increased optical throughput (Jaquinot advantage). Energy-wasting slits are not required in the interferometer because dispersion or filtering is not needed. Instead, a circular optical aperture is commonly used in FTIR systems. The beam area of an FT instrument is usually 75 to 100 times larger than the slit width of a dispersive spectrometer. Thus, more radiation energy is made available. This constitutes a major advantage for many samples or sampling techniques that are energy-limited.

(iii) Internal laser reference (Connes advantage). The use of a helium neon laser as the internal reference in many FTIR systems provides an automatic calibration in an accuracy of better than 0.01 $cm^{-1}$. This eliminates the need for external calibrations.

(iv) Simpler mechanical design. There is only one moving part, the moving mirror, resulting in less wear and better reliability.

(v) Elimination of stray light and emission contributions. The interferometer in FTIR modulates all the frequencies. The unmodulated stray light and sample emissions (if any) are not detected.

(vi) Powerful data station. Modern FTIR spectrometers are usually equipped with a powerful, computerized data system. It can perform a wide variety of data processing tasks such as Fourier transformation, interactive spectral subtraction, baseline correction, smoothing, integration, and library searching.

Although the spectra of many samples can be satisfactorily run on either FTIR or dispersive instruments, FTIR spectrometers are the preferred choice for samples that are energy-limited or when increased sensitivity is desired. A wide range of sampling accessories is available to take advantage of the capabilities of FTIR instruments.

## Sampling Procedures

Solid samples can be handled in a number of ways. A *mull*, or paste is made by grinding a few milligrams of sample in an agate mortar with a drop of a liquid paraffin (e.g., Nujol) or hexachlorobutadiene. The paste is formed into a thin film by compressing it between sodium chloride or potassium bromide plates which are mounted in a suitable sample holder. A complete spectrum of the sample can be obtained from separate mulls in Nujol and hexachlorbutadiene. The C—H band in the former obscure the sample bands in that region but these can be observed in the spectrum of the chlorinated butadiene mull. An alternative procedure is to grind a few milligrams of sample with about 200 mg of dried potassium bromide and to compress the mixture into a thin, homogenous disc using a hydraulic press. Liquid samples can be examined *neat* as a thin film between sodium chloride or potassium bromide plates or as a solution in a cell of path-length 0.1 to 10 mm. Three solvents in widespread use and which are largely transparent in the infrared region, are carbon tetrachloride, chloroform and carbon disulphide. Spectra of the surface of solids and liquid samples can also be obtained either by attenuated total reflectance (ATR) or by diffuse reflectance. The former involves placing the sample in contact with the sides of an optically flat thin prismatic crystal of thallium bromide/iodide (KRS-5) through which radiation from the source is passed. BY directing the beam into one end of the crystal at the correct angle of incidence it can be made to undergo multiple internal reflections before emerging from the other end and passing through the spectrometer to the detector. At each reflection, the beam penetrates the surface of the sample to a depth of a few micrometers. The intensity of the beam is attenuated according to the absorption characteristics of the sample so enabling an absorption spectrum to be recorded. Diffuse reflectance spectra are obtained using a specailly designed curved or plain mirror system to collect the radiation diffusely reflected from the surface of a sample over a wide solid angle and focussing it onto the detector. It is particularly suitable for powdered solid samples, which are simply placed in a small cup, and for liquids or gases adsorbed onto the surface of a non-absorbing powdered substrate such as alkali halide. As little as a few micrograms of material can give an acceptable spectrum.

*Infrared microscopy* combines an optical microscope with an FT-ir spectrometer enabling pico-to femtogram ($10^{-12}$ to $10^{-15}$ g) quantities of substances to be characterized or very small areas of large smaples to be analysed. Beam-condensing optics focus the radiation onto an area of the sample identified using the optical microscope and either reflectance or transmittance spectra can be recorded. The highly-sensitive MCT detector is normally used as its size can be matched to that of the radiation beam to maximize its response.

Gaseous samples require long path length cells to produce absorption bands of reasonable intensity; up to several metres of optical path are obtainable from cells incorporating mirrors which produce multiple reflections. For gc-ir light pipes provide the best sensitivity.

It is possible to obtain in IR spectrum from smaples in many different forms, such as liquid, solid, and gas. However, many materials are opaque to IR radiation and must be dissolved or diluted in a transparent matrix in order to obtain spectra. Alternatively, it is possible to obtain reflectance or emission spectra directly from opaque samples. Some popular sampling techniques and accessories are discussed here.

Liquid cells are used for dilute solutions of solid and liquid samples that are dissolved in relatively IR-transparent solvents. Sampling in solution results in enhanced reproducibility and is often the preferred choice. Unfortunately, no single solvent is transparent through the entire mid IR region. The analyst usually chooses solvents that have transparent windows in the region of interest.

The conventional popular solvents are carbon tetrachloride for the region between 4000 and 1330 $cm^{-1}$ and carbon disulfide for the region between 1330 and 625 $cm^{-1}$. Both solvents are quite toxic, and thus must be handled carefully. One may replace carbon tetrachloride with the less-toxic tetrachloroethylene or methylene chloride and substitute carbon disulfide with n-hexane or n-heptane. Polar solvents such as water and alcohols are seldom used because they absorb strongly in the mild IR range and react with alkali-metal halides, such as NaCl, commonly used for cell windows.

Acquiring acceptable IR spectra of aqueous samples requires use of special types of liquid cells such as thin cells of $BaF_2$, AgCl, or KRS-5 (a mixed thallium bromide-thallium iodide). Aqueous solution measurements can also be accomplished with attenuated total reflectance (ATR) accessories, which are discussed later in this chapter.

Typically, solutions of 0.05 to 10% in concentration are handled in IR cells of 0.1 to 1 mm in thickness. Concentration of 10% and cell path length of 0.1 mm represent one practical combination. In a double-beam spectrometer, a compensating cell is filled with pure solvent and placed in the reference beam. In the single-beam FT instrument, the solvent bands are mostly removed by obtaining the difference spectra through subtraction of solvent spectra from sample spectra. Both fixed-thickness and variable-thickness liquid cells are available commercially. They normally consist of metal frame plates, IR-transmitting windows, and gaskets that determine the path length of the cells.

Salt plates of IR-transmitting materials can be used for semivolatile and nonvolatile liquid samples. Sodium chloride disks are the most popular and economical choice for nonaqueous liquids. Silver chloride or barium fluoride plates may be used for samples that dissolve or react with NaCl plates. A drop of the neat sample is squeezed between two salt plates to form a film of approximately 0.01 mm in thickness. The plates can be held together by capillary attraction, or they may be clamped in a screw-tightened holder or pressed to form a good contact in a press fit O-ring supported holder. It is also possible to place a film of samples on salt plates by melting a relatively low-melting solid and squeezing it between two plates. Sodium chloride slat plates can usually be cleaned with dry methylene chloride or acetone. This smear technique is one of the simplest ways to obtain IR spectra.

Thin films of nonvolatile liquids or solids can be deposited on an IR-transmitting salt plate by solvent evaporation. The sample is first dissolved in a reasonably volatile solvent. A few drops of the resulting solution are placed on the plate. After evaporating off the solvent, a thin film of sample is obtained for subsequent spectra acquisition.

Disposable IR cards have been developed recently by 3M to accommodate samples that are liquids, are soluble in reasonably volatile solvents, or can be smeared on flat surfaces. The cards are made up of a cardboard holder containing a circular IR-transmitting window made of a microporous substrate (polytetrafluoroethylene substrate for 4000 to 1300 $cm^{-1}$ or polyethylene substrate for 1600 to 400 $cm^{-1}$). Samples are generally applied to the cards by the techniques used for salt plates. The substrate hands can be subtracted from the sample spectra. Besides the convenience, the disposable IR cards are nonhygroscopic, and thus can handle water-containing samples.

Pellets are used for solid samples that are difficult to melt or dissolve in any suitable IR-transmitting solvents. The sample (0.5 to 1.0 mg) is finely ground and intimately mixed with approximately 100 mg of dry potassium bromide (or other alkali halide) powder. Grinding and mixing can be done with an agate mortar and pestel, a vibrating ball mill (Wig-L-Bug from Cres-cent Dental Manufacturing), or lyophilization. The mixture is then pressed into a transparent disk in an evacuable die at sufficiently high pressure. Suitable KBr disks or pellets can often be made using a simpler device such as a Mini-Press. To minimize band distortion due to scattering of radiation, the sample should be ground to particles of 2 μm (the low end of the radiation wavelength) or less in size. The IR spectra produced by the pellet technique often exhibit bands at 2450 $cm^{-1}$ and 1640 $cm^{-1}$ due to absorbed moisture.

Mulls are used as alternatives for pellets. The sample (1 to 5 mg) is ground with a mulling agent (1 to 2 drops) to give a two-phase mixture that has a consistency similar to toothpaste. This mull is pressed between two IR-transmitting plates to form a thin film. The common mulling agents include mineral oil or Nujol (a high-boiling hydrocarbon oil), Fluorolube (a chlorofluoro-carbon polymer), and hexachlorobutadiene. To obtain a full IR spectrum that is free of mulling agent bends, the use of multiple mulls (such as Nujol and Fluorolube) is generally required. Thorough mixing and reduction of sample particles of 2 μm or less in size are very important in obtaining a satisfactory spectrum.

Gas cells can be used to examine gases or low-boiling liquids. These cells consist of a glass or metal body, two IR-transparent end windows, and vales for filling gas from external sources. They provide vacuum-tight light paths from a few centimeters to 120 m. Longer path lengths are obtained by reflecting the IR beam repeatedly through the sample using internal mirrors located at the ends of the cell. Sample gas pressure required to produce reasonable spectra depends on the sample absorbance and the cell's path length. Typically, a good spectrum can be acquired at a partial pressure of 50 torr in a 10–cm cell. Analysis of multicomponent gas samples at parts-per-billion levels can be successfully performed.

Microsampling accessories such as microcells, microcavity cells, and micropellet dies are used to examine microquantities of liquids (down to 0.5 μL) and solids (down to 10 μg). Beam condensing devices are often used to reduce the beam size at the sampling point. Extra practice is recommended for performing this type of microanalysis.

Attenuated total reflectance (ATR) accessories are especially useful for obtaining IR spectra of difficult samples that cannot be readily examined by the normal transmission method. They are suitable for studying thick or highly absorbing solid and liquid materials, including films, coatings, powders, threads, adhesives, polymers, and aqueous samples. ATR requires little or no sample preparation for most samples and is one of the most versatile sampling techniques.

ATR occurs when a beam of radiation enters from a more-dense (with a higher refractive index) into a less-dense medium (with a lower refractive index). The fraction of the incident beam reflected increases when the angle of incidence increases. All incident radiation is completely reflected at the interface when the angle of incidence is greater than the critical angle (a function of refractive index). The beam penetrates a very short distance beyond the interface and into the less dense medium before the complete reflection occurs. This penetration is called the evanescent wave and typically is at a depth of a few micrometers (μm). Its intensity is reduce (attenuated) by the sample in regions of the IR spectrum where the sample absorbs. Fig. 20 illustrates the basic ATR principles.

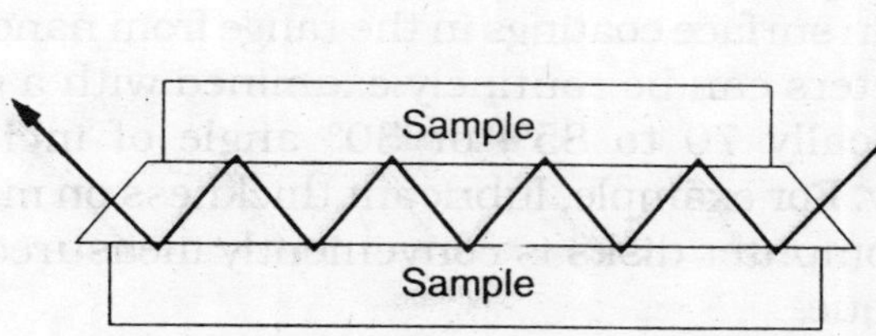


**Fig. 20. Schematic representation of multiple internal reflection effect in Attenuated Total Reflectance (ATR).**

The sample is normally placed in close contact with a more-dense, high-refractive-index crystal such as zinc selenide, thallium bromide-thallium iodide (KRS-5), or germanium. The IR beam is directed onto the beveled edge of the ATR crystal and internally reflected through the crystal with a single or multiple reflections. Both the number of reflections and the penetration depth decrease with increasing angle of incidence. For a given angle, the higher length-to-thickness ratio of the ATR crystal give higher numbers of reflections. A variety of types of ATR accessories are available, such as 25 to 75° vertical variable-angle ATR, horizontal ATR, and Spectra-Tech Cylindrical Internal Reflectance Cell for Liquid Evaluation (CIRCLE®) cell.

The resulting ATR-IR spectrum resembles the conventional IR spectrum, but with some differences: The absorption band positions are identical in the two spectra, but the relative intensities of corresponding bands are different. Although ATR spectra can be obtained using either dispersive or FT instruments, FTIR spectrometers permit higher-quality spectra to be obtained in this energy-limited situation.

Specular reflectance provides a nondestructive method for measuring thin coatings on selective, smooth substrates without sample preparation. It basically involves a mirrorlike reflection and produces reflection measurements for a reflective material, or a reflection-absorption spectrum for the surface film on a reflective surface. Thin surface coatings in the range from nanometers to micrometers can be routinely examined with a grazing angle (typically 70 to 85°) or 30° angle of incidence, respectively. For example, lubricant thickness on magnetic media or computer disks is conveniently measured using this technique.

Diffuse reflectance technique is mainly used for acquiring IR spectra of powders and rough surface solids such as coal, paper, and cloth. It can be used as an alternative to pressed-pellet or mull techniques. IR radiation is focused onto the surface of a solid sample in a cup and results in two types of reflections: sepcular reflectance, which directly reflects off the surface and has equal angles of incidence and reflectance, and diffuse reflectance, which penetrates into the sample, then scatters in all directions. Special reflection accessories are designed to collect and refocus the resulting diffusely scattered light by large ellipsoidal mirrors, while minimizing or eliminating the specular reflectance, which complicates and distorts the IR spectra. This energy-limited technique was not popular until the advent of FTIR instruments. This technique is often called diffuse reflectance infrared Fourier transform spectroscopy (DRIFTS).

The sample can be analyzed either directly in bulk form or as dispersions in IR-transparent matrices such as KBr and KCl. Dilution of analyte in a nonabsorbing matrix increases the proportion of diffuse reflectance in all the light reflected. Typically the solid sample is diluted homogeneously to 5 to 10% by weight in KBr. The spectra of diluted samples are similar to those obtained from pellets when plotted in units such as log $1/R$ ($R$ is the reflectance) of Kubelka-Munk units. The Kubelka-Munk format relates sample concentration to diffuse reflectance and applies a scattering factor.

Photoacoustic spectroscopy (PAS) is a useful extension of IR spectroscopy and suitable for examining highly absorbing samples that are difficult to analyze by conventional IR techniques. The size and shape of the sample are not critical. PAS spectra can be obtained with minimal sample preparation and without physical alteration from a wide variety of sample such as powders, polymer pellets, viscous glues, single crystals, and single fibers.

Typically, the modulated IR radiation from an FTIR interferometer is focused on a sample placed in a small cup inside a small chamber containing an IR-transparent gas such as helium or nitrogen. IR radiation absorbed by the sample converts into heat inside the sample. The heat diffuses to the sample surface, then into the surrounding gas atmosphere, and causes expansion of a boundary layer or gas next to the sample surface. Thus, the modulate IR radiation produces intermittent thermal expansion of the boundary layer and generates pressure waves. A sensitive microphone is used to detect the resulting photacoustic signal.

PAS spectra are generally similar to conventional IR spectra except for some minor differences: Absorbance peaks appear at the same frequency locations, but truncation of strong absorbance bands due to photoacoustic signal saturation is often observed. However, the presence of such truncated bands does not limit the practical use of PAS. Spectral search against standard commercial spectra libraries can be satisfactorily performed. FTIR PAS technique also offers a unique capability for examining samples at various depths from 1 to 20 μm. The acoustic frequencies depend on the modulated frequency of source. The slower the modulation frequency, the greater depth of penetration. Thus, sample such as multilayer polymers can be studied at various depths by simply varying the scan speed of the FTIR spectrometer.

Emission spectroscopy is another technique used with difficult samples such as thin coatings and opaque materials. The sample is normally heated to an elevated temperature, emitting enough energy to be detected. The sample acts as the radiation source, so the normal IR source is turned off. The ability of FTIR instruments to obtain spectra from weak signals makes it possible to study emission in the infrared region, even when the sample is at low temperatures such as 50 at 100°C. Emission spectral bands occur at the same frequencies as absorption bands. The spectra from thick samples can be complicated when radiation from the interior of the sample is self-absorbed by the outer part of the sample.

Infrared microspectroscopy has become a popular technique for analyzing difficult or small samples such as trace contaminants in semiconductor processing, multilayer laminates, surface defects, and forensic samples. Infrared microscopes are energy-inefficient accessories that require the signal-to-noise advantages of FTIR to obtain spectra from submilligram samples. Using a liquid nitrogen cooled mercury cadmium telluride (MCT) detector, samples in the size range of 10 μm can be examined on IR microscopes.

The primary advantages of the IR microscope relate not only to its improved optical and mechanical design, but also to its manipulative capability. In many cases, the major problem in microsampling is focusing the spectrometer beam on the sample. The computerized/ motorized control of microscope functions of IR microscope instruments permit these extremely small samples to be moved in the field of view to isolate the portion from which spectra are obtained.

Fiberoptic accessories deliver unique flexibility and diversity in sampling. They are particularly useful in acquiring IR spectra when samples are situated in a remote location or when the unusual size or shape of samples prevents them from fitting well in a standard sample compartment. Many analyses in hazardous or process environments used these devices.

Fiberoptic sample probes or flow cells are coupled to standard FTIR spectrometers with two fiberoptic cables and an optic interface that transfers IR radiation from spectrometer to fiberoptic cables. A variety of probes are available for ATR, specular reflectance, diffuse reflectance, and transmittance measurements. Chalcogenide (GeAsSeTe), a mid IR-transmitting material in the range of 4000 to 900 $cm^{-1}$, was recently developed by Spectra-Tech and used to make the fiberoptic cables.

## Hyphenated Methods Involving Infrared

Gas chromatography/Fourier transform infrared (GC/ FTIR) spectroscopy is a technique that uses a gas chromatography to separate the components of sample mixtures and an FTIR spectrometer to provide identification or structural information on these components. The real potential of GC-IR instrumentation was not widely used until the fast-scanning, sensitive FTIR spectrometers became available commercially.

The most commonly used GC/FTIR interface is a light pipe flow cell. The light pipe is typically a piece of glass tubing 10 to 20 cm long, approximately 1 mm inside diameter, gold coated on the inside, with IR-transmitting windows on each end. This design provides a long path length and low dead volume (90 and 300 μL), resulting in high IR absorbance with minimal peak broadening. The light pipe is connected to the effluent port of the gas chromatograph by a heated transfer line. The gas flow assembly can be heated up to 350°C to prevent sample components from condensing onto the light pipe and transfer line. Fig. 21 illustrates the optical design of a GC/FTIR interface.

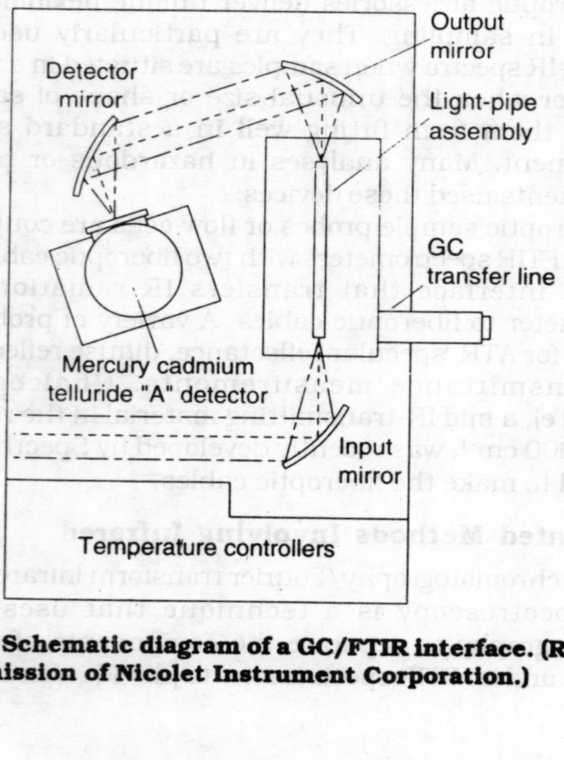


**Fig. 21. Schematic diagram of a GC/FTIR interface. (Reprinted by permission of Nicolet Instrument Corporation.)**

Eluents from a capillary gas chromatograph flow through the transfer line into the light pipe, where the IR spectra are acquired in real time with a rate up to 20 spectra per second. The light pipe GC/FTIR offers nanogram-level sensitivity. Typically, a usable spectrum can be obtained from 5 to 20 ng of component compound. The flow emerging from the chromatograph is often split between the light-pipe and a conventional GC detector (flame ionization, thermal conductivity, or mass spectrometer). This permits the simultaneous generation of a normal chromatogram and the corresponding IR spectra for each chromatographic peak. Alternatively, the total flow after the light pipe can be routed into a conventional detector to provide in-line detection by a flame ionization or mass spectrometer detector. In fact, the combination of a gas chromatograph with and FTIR an mass spectrometer (GC/FTIR/MS) is available commercially.

Although common GC/FTIR spectroscopy is not as sensitive as gas chromatography/mass spectrometry (GC/MS), GC/FTIR offers a major advantage over GC/MS: the ability to identify structural isomers. In addition, the sensitivity of GC/FTIR can be further improved by matrix isolation or direct deposition techniques.

Gas chromatography/matrix isolation/Fourier transform infrared (GC/MI/FTIR) spectroscopy provides subnanogram sensitivity, but is a very expensive technique. The helium carrier gas of a gas chromatograph is mixed with a small amount or argon. While argon is condensed in a track of 300 μm width on a rotating circular gold-coated metal disk cooled at 12°K, the helium gas is evacuated by pumping. The components separated by the chromatograph are dissolved and trapped in the argon matrix. After the GC run is completed, the argon track is rotated into the IR beam and the reflection-absorption IR spectra are obtained for each component on the cooled-surface. Cryogenic temperatures are maintained while the spectra are acquired.

In GC/MI/FTIR, the components are isolated in small areas. Because IR measurements are not made in real time, these components can be held in the IR beam for longer periods, allowing improved single-to-noise ratios through averaging of multiple scans. Thus, GC/MI/FTIR offers significant sensitivity improvement over light-pipe GC/FTIR.

Gas chromatography/direct deposition/Fourier infrared (GC/DD/FTIR) spectroscopy is another sensitive technique that permits a usable spectrum to be obtained with 100 pg of component compound. The separated components are directly deposited in a track of 100 μm width on a liquid-nitrogen-cooled, IR-transmitting disk such as zinc selenide. Transmission IR spectra can be taken in real time. Alternatively, the isolated components can be repositioned in the IR beam after the run is completed to gain the multiscan signal-averaging advantage.

High-performance liquid chromatography/Fourier transform infrared (HPLC/FTIR) spectroscopy uses the same approach as the GC/DD/FTIR to eliminate the mobile phase and gain satisfactory sensitivity.

Conventional flow cells for HPLC chromatograph generally do not provide adequate sensitivity, due to the IR absorption of all HPLC mobile phase. Instead, an HPLC/FTIR interface using the direct deposition technique has been designed and delivers subnanogram sensitivity (1). The interface consists of two concentric fused silica tubes of different internal diameters. While the eluate stream from a HPLC column flows through the inner tube, a sheath of hated gas (helium or air) passes through the outer tube. The nebulized spray is directed to a rotating sample collection disk onto which the component compounds are deposited. During the process the mobile phase is evaporated. The resulting depositions of sample components can then be positioned in the IR beam and their IR spectra collected.

Other techniques involving IR spectrometers, including

supercritical fluid chromatography/Fourier transform infrared (SFC/FTIR) spectroscopy, thermogravimetry/Fourier transform infrared (TGA/FTIR) spectroscopy and, gas chromatography/Fourier transform infrared/mass spectrometry (GC/FTIR/MS) have also become available commercially. They generally use more cost-effective flow-through interfaces.

## Analytical Information

### Qualitative

The combination of the fundamental vibrations or rotations of various functional groups and the subtle interactions of these functional groups with other atoms of the molecule results in the unique, generally complex IR spectrum for each individual compound. IR spectroscopy is mainly used in two ways: structural elucidation and compound identification.

### Structural Elucidation

Because of complex interactions of atoms within the molecule, IR absorption of the functional groups may vary over a wide range. However, it has been found that many functional groups give characteristic IR absorption at specific, narrow frequency ranges regardless of their relationship with the rest of the molecule. Generalized tables of the positions and relative intensities of absorption bands (Fig. 15.13) have been established and used to determine the functional groups present or absent in a molecule. The *CRC Handbook of Chemistry and Physics* (2), Silverstein, Bassler, and Morrill's book (3), and a number of other publications all contain useful correlation charts.

Multiple functional groups may absorb at one particular frequency range, but a functional group often give rise to multiple-characteristic absorption. Thus, the spectral interpretations should not be confined to one or two bands and the whole spectrum should examined. To confirm or better elucidate the structure of an unknown substance, other analytical information provided by nuclear magnetic resonance (NMR), mass spectrometry (MS), or other chemical analysis should also be used where possible. For systematic evaluation, the IR spectrum is commonly divided into three regions.

The Functional Group Region, 4000 to 1300 cm$^{-1}$. The appearance of strong absorption hands in the region 4000 to 2500 cm$^{-1}$ usually comes from stretching vibrations, between hydrogen and some other atoms with a mass of 19 or less. The O–H and N–H stretching frequencies fall in the 3700 to 2500 cm$^{-1}$ region, with various intensities. Hydrogen bonding has a significant influence on the peak shape and intensity, generally causing peak broadening and shifts in absorption to lower frequencies. The C–H stretching bands occur in the region of 3300 to 2800 cm$^{-1}$. The acetylenic C-H exhibits strong absorption at about 3300 cm$^{-1}$. Alkene and aromatic C—H stretch vibrations absorb at 3100 to 3000 cm$^{-1}$. Most aliphatic (saturated) C-H stretching bands occur at 3000 to 2850 cm$^{-1}$, with generally prominent intensities that are proportional to the number of C-H bonds. Aldehydes often show two sharp aldehydic C-H stretching absorption bands at 2900 to 2700 cm$^{-1}$.

The absorption bands at the 2700 to 1850 cm$^{-1}$ rgion usually come only from triple bonds and other limited types of functional groups, such as C ≡ C at 2260 to 2100 cm$^{-1}$, C ≡ N at 2260 to 2220 cm$^{-1}$, diazonium salt $-N^{+} \equiv N$ at approximately 2260 cm$^{-1}$, allenes C=C=C at 2000 to 1900 cm$^{-1}$, S-H at 2600 to 2550 cm$^{-1}$, P-H at 2440 to 2275 cm$^{-1}$, Si-H at 2250 to 2100 cm$^{-1}$.

The 1950 to 1450 cm$^{-1}$ region exhibits IR absorption from a wide variety of double-bonded functional groups. Almost all the carbonyl C=O stretching bands are strong and occur at 1870 to 1550 cm$^{-1}$. Acid chlorides and acid anhydrides give rise to IR bands at 1850 to 1750 cm$^{-1}$. Whereas ketones aldehydes, carboxylic acids, amides, and esters generally show IR absorption at 1750 to 1650 cm$^{-1}$, carboxylate ions usually display stretching bands at 1610 to 1550 and 1420 to 1300 $cm^{-1}$. Conjugation, ring size, hydrogen bonding, and steric and electronic effects often result in significant shifts in absorption frequencies. Nonconjugated aliphatic C=C and C=N have absorption bands at 1690 to 1620 $cm^{-1}$, with variable intensities. Aromatic compounds contain delocalized $\pi$ electrons from the resonance-stabilized double bonds, showing skeletal vibrations (including C-C stretchings within the ring) in the 1650 to 1440 $cm^{-1}$ region and weak combination and overtone bands in the 2000 to 1650 $cm^{-1}$ region. Valuable information about the substitution pattern on an aromatic ring can be obtained by careful examination of absorption bands in these two regions. Molecules containing $NO_2$ groups, such as nitro compounds, nitrates, and nitramines, commonly exhibit asymmetric and symmetric stretching vibrations of the $NO_2$ groups at 1660 to 1500 and 1390 to 1260 $cm^{-1}$ regions.

The Fingerprint Region, 1300 to 910 $cm^{-1}$ Absorptions in this region include the contributions from complex interacting vibrations, giving rise to the generally unique fingerprint for each compound. A good match between the IR spectra of two compounds in all frequency ranges, particularly in the fingerprint region, strongly indicates that they have the same molecular structures.

Detailed interpretation of IR bands in this region is difficult. However, some assignments of bands in the fingerprint region to a few important vibrational frequencies of functional groups can be done when IR absorptions in other regions are correlated together. For example, esters not only show their carbonyl C=O stretch at 1750 to 1735 $cm^{-1}$, but also exhibit their characteristic absorption at 1300 to 1000 $cm^{-1}$ from the the couplings of C-O and C-C stretches.

Detailed interpretation of IR bands in this region is difficult. However, some assignments of bands in the fingerprint region to a few important vibrational frequencies of functional groups can be done when IR absorptions in other regions are correlated together. For example, esters not only show their carbonyl C=O stretch at 1750 to 1735 $cm^{-1}$, but also exhibit their characteristic absorption at 1300 to 100 $cm^{-1}$ from the couplings of C-O and C-C stretches.

The Aromatic Region, 910 to 650 $cm^{-1}$. The IR bands in this region do not necessarily come from the aromatic compounds, but the absence of strong absorption in the 910 to 650 $cm^{-1}$ region usually indicates the lack of aromatic characters. The out-of-plane bending of ring C-H bonds of aromatic and heteroaromatic compounds gives rise to strong IR bands in the range between 910 and 650 $cm^{-1}$. As previously stated, certain nonaromatic molecules such as amines and amides can also contribute absorption in this region.

## Compound Identification

Since the IR spectrum of every molecule is unique, one of the most positive identification methods of an organic compounds is to find a reference IR spectrum that matches that of the unknown compound.

A large number of reference spectra for vapor and condensed phases are available in printed and electronic formats. The spectral libraries compiled by Sadtler and Aldrich are some of the most popular collections. In addition, spectral databases are often compiled according to application areas such as forensics, biochemicals, and polymer. Computerized search programs can facilitate the matching process. In many cases where exact match to the spectrum of an unknown material cannot be found, these programs usually list the reference compounds that match the unknown spectrum most closely. This information is useful in narrowing the search. When it is combined with the data from other analysis such as NMR or mass spectrometry, a positive identification or high-confidence level tentative identification can often be achieved.

The value of infrared spectrometry as a means of indentifaction of unknown compounds and to investigate structural features is immense. Spectra are used in an empirical manner by comparison of samples with known materials and by reference to charts of group frequencies. A simplified correlation chart is shown in table 9.8. The interpretation of infrared spectra is best considered by discussing the prominent features of a representative series of compounds.

The spectrum of a saturated hydrocarbon, *n*-decane is shown in fig. 22. Four prominent bands arise from C—H stretching and bending vibrations: asymmetric and symmetric C—H stretching (~2900 cm$^{-1}$), $CH_2$ bending and asymmetric $CH_3$ bending (~1460 cm$^{-1}$), symmetric $CH_3$ binding (~1380 cm$^{-1}$) and rocking of all the $CH_2$ groups in the chain (~720 cm$^{-1}$). The position of the latter depends upon chain length and is not observed if there are fewer than four methylene groups together. The C—H stretching bands may be further resolved into components due to $CH_2$ and $CH_3$ groups. It should be noted that no bands are observed above 3000 cm$^{-1}$ indicating the absence of an aromatic, olefinic or acetylenic structure.

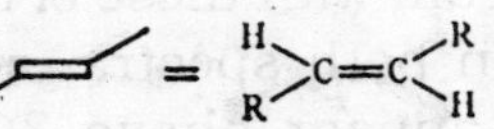

and so on.

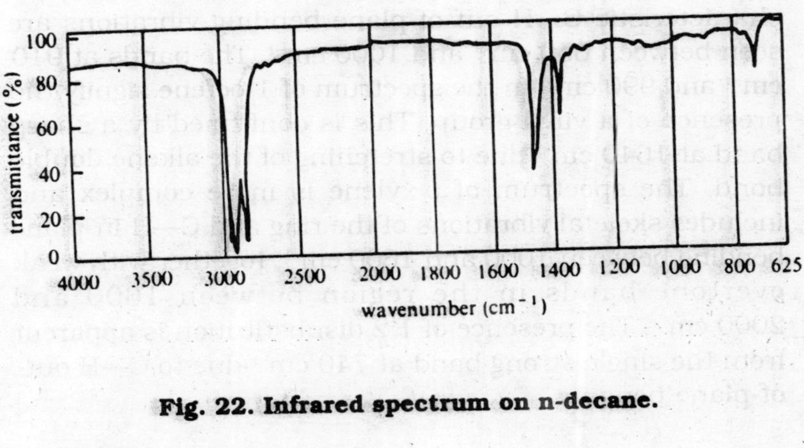


**Fig. 22. Infrared spectrum on n-decane.**

The charts are arranged in order of increasing element content. Correlations given in one chart (e.g., those for $CH_2$ and $CH_3$) are not repeated in subsequent charts. The frequency limits within which the band of a particular grouping is usually found are indicated by the black strips and extensions of the range to include unusual examples are shown as thin lines, e.g., relative intensities are given in a very approximate fashion. Both the position and the intensity of some absorption are dependent on state, solvent, etc. and the actual frequency quoted is that most commonly observed.

## Abbreviations

v = stretch ($\upsilon$ is also for quoting an absorption frequency, e.g. $\upsilon_{max}$ 1700 cm$^{-1}$); $\delta$ = in-plane bend; $\gamma$ = out-of-plane bend; $\rho$ = rock; $\omega$ = wag; $\tau$ = twist; sym = symmetrical; antisym = antisymmetrically (asymmetric); var. = variable (in frequency); *br* = broad; *w* = weak (in absorption); *m* = medium; *s* = strong; *vw* = very weak; *vs* = very strong; *w-m* = weak to medium (i.e., typical intensity range for that particular band).

Contrast this spectrum with those of *o*-xylene, fig. 23, and 1-octene, fig. 24. In both spectra, additional C—H stretching vibrations appear above 3000 cm$^{-1}$, and characteristic C—H out-of-plane bending vibrations are seen between 650 cm$^{-1}$ and 1000 cm$^{-1}$. The bands at 910 cm$^{-1}$ and 990 cm$^{-1}$ in the spectrum of 1-octene signify the presence of a vinyl group. This is confirmed by a sharp band at 1640 cm$^{-1}$ due to stretching of the alkene double bond. The spectrum of *o*-xylene is more complex and includes skeletal vibrations of the ring and C—H in-plane bending between 1000 and 1600 cm$^{-1}$, together with weak overtone bands in the region between 1600 and 2000 cm$^{-1}$. The presence of 1,2 disubstitution is apparent from the single strong band at 740 cm$^{-1}$ due to C—H out-of-plane bending.

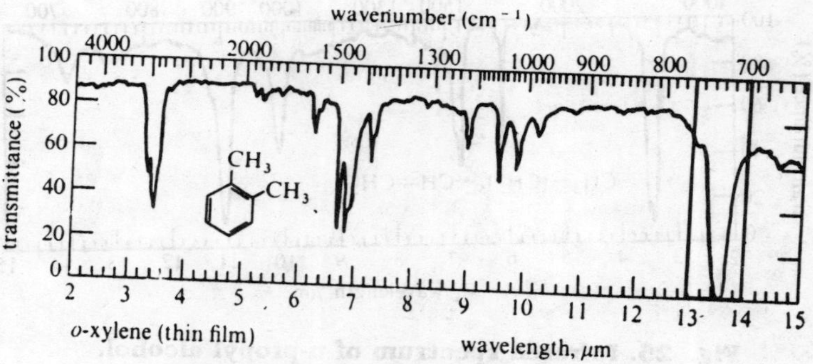


**Fig. 23. Infrared spectrum of o-xylene.**

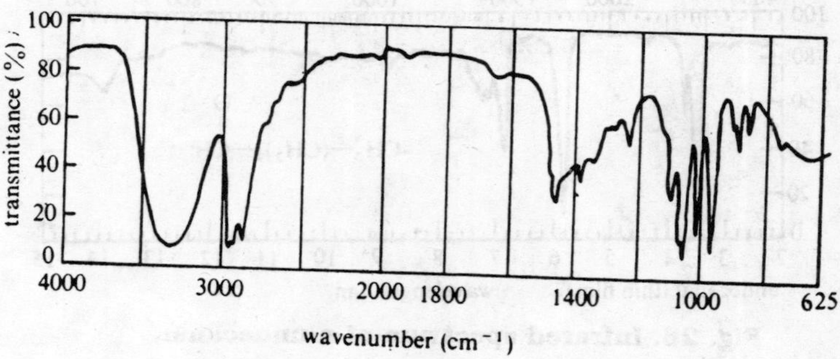


**Fig. 24. Infrared spectrum of 1-octene.**

Compare, also, the spectrum of *n*-decane with those of *n*-propyl alcohol, fig. 25 and *n*-undecanal, fig. 26. The spectrum of the alcohol is characterized by vibrations due to the O—H and C—O groups. The O—H stretching band at ~ 3350 $cm^{-1}$ is broadened by hydrogen bonding while the strong band at ~ 1070 $cm^{-1}$ is due to a complex C—O stretching vibration. In the aldehyde a strong band assigned to stretching of the carbonyl group appears at ~ 1730 $cm^{-1}$ and a sharp peak due to stretching of the aldehydic C—H (~ 2720 $cm^{-1}$) serves to differentiate it from other types of carbonyl compound.

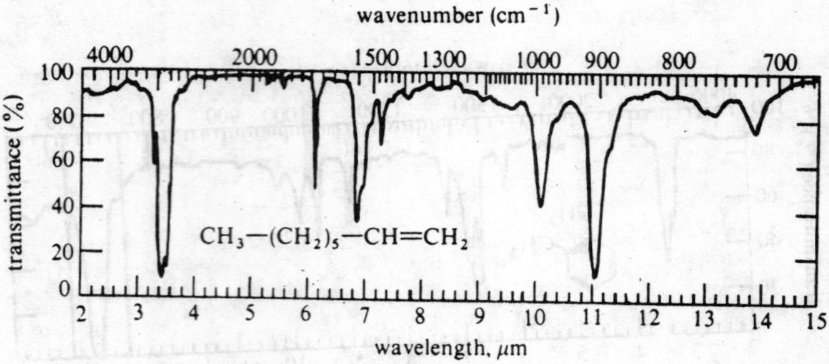


**Fig. 25. Infrared spectrum of n-propyl alcohol.**

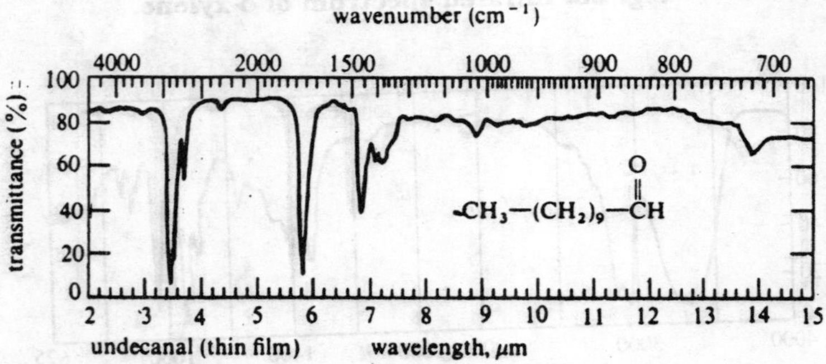


**Fig. 26. Infrared spectrum of n-undecanal.**

The spectrum of a carboxylic acid can be compared with that of an ester in figs. 27 and 28. Such acids exist primarily as dimes, and the strongly hydrogen bonded structures give rise to a very broad O—H stretching band between 2500–3500 $cm^{-1}$ and another at around 940 $cm^{-1}$ due to O—H out-of-plane bending. An intense carbonyl stretching band occurs near 1710 $cm^{-1}$, similar to that of an aliphatic ketone because of hydrogen bonding, and the C—O strech appears as a broad absorption between 1200 and 1300 $cm^{-1}$. The ester can be distinguished by the absence of O—H bands around 3000 and 940 $cm^{-1}$ and the shift of the carbonyl band to ~1740 $cm^{-1}$ caused by replacement of the acidic proton with an alkyl group. (N.B. carbonyl absorption in a monomer acid occurs at ~1760 cm$^{-1}$). Furthermore the C—O stretching band between 1200 and 1300 cm$^{-1}$ is considerably intensified. It should be noted that in the spectra of acids the C—H stretching band is frequently observed as a sharp peak above the broad O—H band.

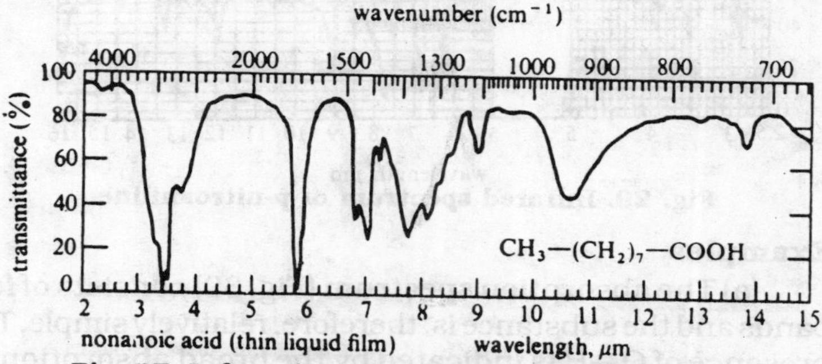


**Fig. 27. Infrared spectrum of n-nonanoic acid.**

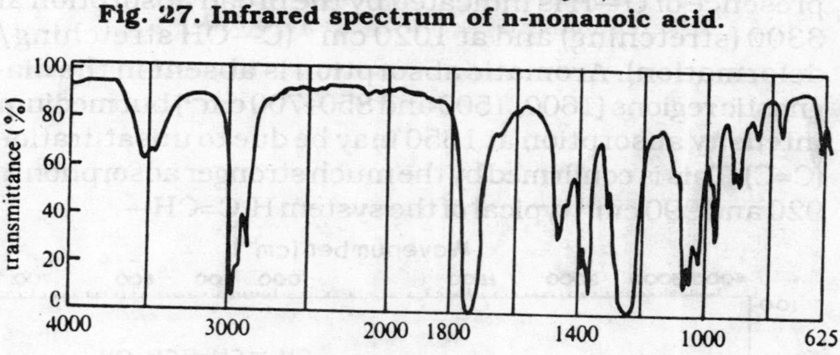


**Fig. 28. Infrared spectrum of n-propyl acetate.**

The final spectrum, fig. 29, is that of *p*-nitroaniline. Although complex, it is immediately recognizable as being aromatic (C—H above 3000 cm$^{-1}$, skeletal and overtone bands between 1400 and 2000 cm$^{-1}$ and a C—H out-of-plane bending vibrations at 835 cm$^{-1}$ indicating *p*-substitution). Asymmetric and symmetric N=O stretching bands are very prominent at 1475 and 1310 cm$^{-1}$, whilst evidence for an amine group is given by the doublet N—H stretching band near 3400 cm$^{-1}$ and the in-plane N—H bending band near 1600 cm$^{-1}$. A strong C—N stretching vibration overlaps the symmetric N=O band near 1300 cm$^{-1}$.

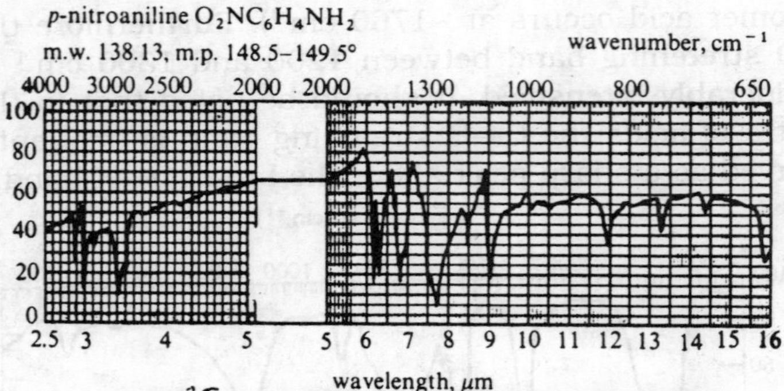


**Fig. 29. Infrared spectrum of p-nitroaniline.**

## Examples

(a) The absorption spectrum (Fig. 30) consists of few bands and the substance is, therefore, relatively simple. The presence of O—H is indicated by the broad absorption at 3300 (stretching) and at 1020 $cm^{-1}$ (C—OH stretching/deformation). Aromatic absorption is absent in the diagnostic regions (1600, 1500 and 850–700 $cm^{-1}$) but medium intensity absorption at 1650 may be due to unsaturation (C=C). This is confirmed by the much stronger adsorption at 920 and 990 $cm^{-1}$ typical of the system $H_2C$=CH—.

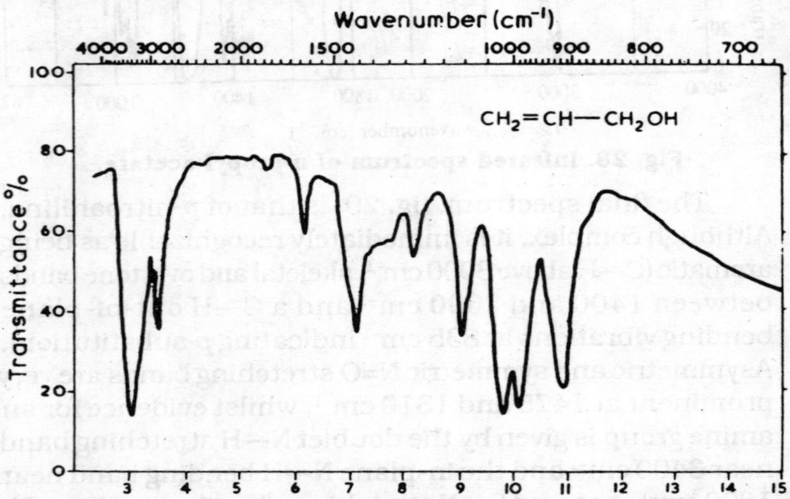


**Fig. 30. Infrared spectrum of allyl alcohol.**

(b) The presence of the hydroxyl group is indicated immediately by the absorption at 3300 and 1020 cm$^{-1}$ (Fig. 31). This region also shows absorption due to C—H in HC=C and $CH_2$ at 3010 and 2800 cm$^{-1}$ respectively. That the unsaturation is of an aromatic nature is evident from the characteristic absorption in the regions 1600, 1500 and 850–700 cm$^{-1}$. Aromatic patterns are also shown between 2000 and 1700 cm$^{-1}$. The hydroxyl group cannot be attached directly to the aromatic system otherwise enhancement of the absorption at 1600 cm$^{-1}$ would be noticeable Aromatic absorption at 695 and 735 cm$^{-1}$ indicates monsubstitution of the benzene ring.

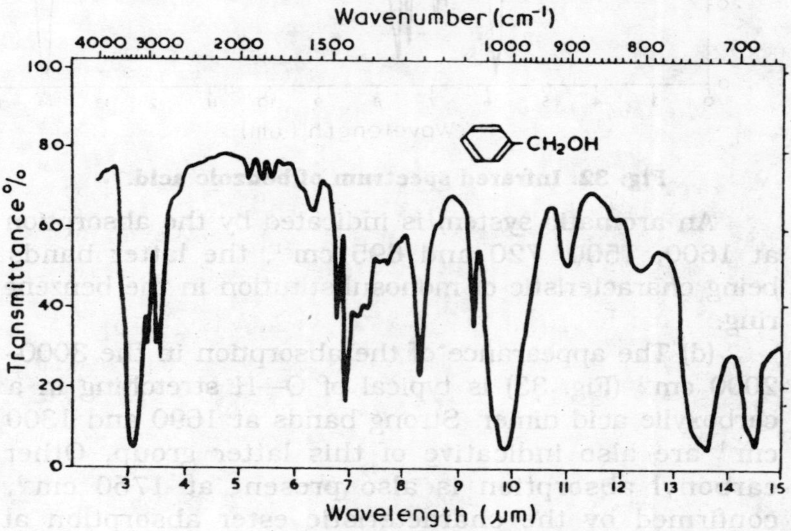


**Fig. 31. Infrared spectrum of benzyl alcohol.**

(c) The broad absorption band at 3000–2500 cm$^{-1}$ (Fig. 32) indicates hydrogen-bonded O–H of —COOH. Hydrochlorides of organic bases give somewhat similar absorption but the carbonyl group is more likely in view of the strong carbonyl absorption at 1690 cm$^{-1}$. Confirmation is given by the bands at 1280 and 1310 cm$^{-1}$. (—C—O—) and by the characteristic broad band at 940 cm$^{-1}$ due to bonded O—H deformation in dimetric acids.

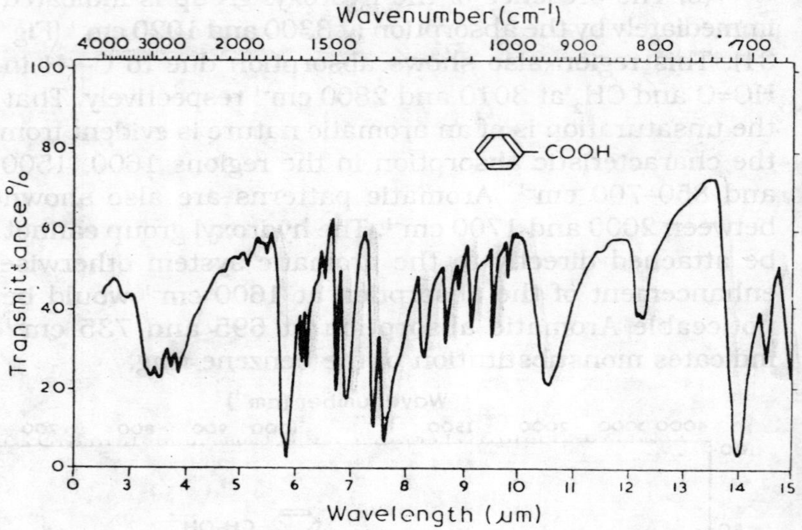


**Fig. 32. Infrared spectrum of benzoic acid.**

An aromatic system is indicated by the absorption at 1600, 1500, 720 and 695 $cm^{-1}$, the latter bands being characteristic of monosubstitution in the benzene ring.

(d) The appearance of the absorption in the 3000–2000 $cm^{-1}$ (Fig. 33) is typical of O—H stretching in a carboxylic acid dimer. Strong bands at 1690 and 1300 $cm^{-1}$ are also indicative of this latter group. Other carbonyl absorption is also present at 1750 $cm^{-1}$, confirmed by the characteristic ester absorption at 1180 and 1220 $cm^{-1}$.

The compound is aromatic as shown by the bands at 1600, 1490 and 850–700 $cm^{-1}$. The absorption at 1600 $cm^{-1}$ is very strong for an aromatic compound but it can be explained as being enhanced by the presence of polar substituents.

(e) Aliphatic C—H is present only in small amount because of the very weak absorption at about 3000 $cm^{-1}$(Fig. 34). The band at 3200 $cm^{-1}$ may be due to hydroxyl but is probably (from the intensity and appearance), due to NH stretching. An aromatic system is confirmed by the bands at 1580, 1500 cm$^{-1}$, and by those in the 850–700 cm$^{-1}$ region.

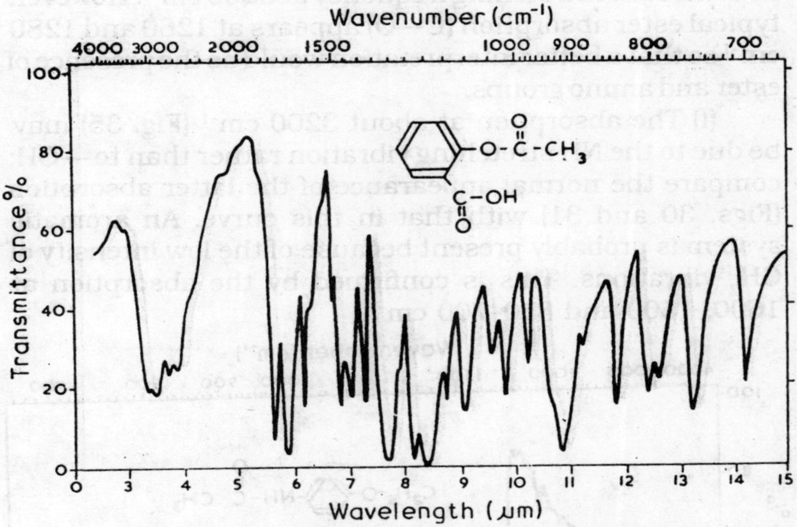


**Fig. 33. Infrared spectrum of acetylsalicylic acid.**

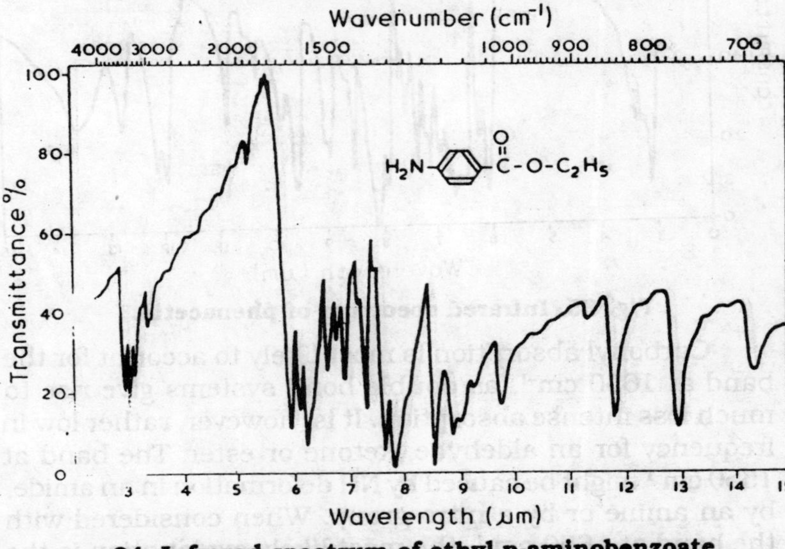


**34. Infrared spectrum of ethyl p-aminobenzoate.**

Strong carbonyl absorption at 1680 cm$^{-1}$ might easily be interpreted as amide —C=O with the band at 1580 cm$^{-1}$ as the second peak for amide absorption. This is in agreement with the N—H stretching frequency at 3200 cm$^{-1}$. However, typical ester absorption (C—O) appears at 1260 and 1280 cm$^{-1}$ so that a better interpretation would be the presence of ester and amino groups.

(f) The absorption at about 3200 cm$^{-1}$ (Fig. 35) may be due to the NH stretching vibration rather than to —OH; compare the normal appearance of the latter absorption (Figs. 30 and 31) with that in this curve. An aromatic system is probably present because of the low intensity of $CH_2$ vibrations. This is confirmed by the absorption at 1600, 1500 and 850–700 cm$^{-1}$.

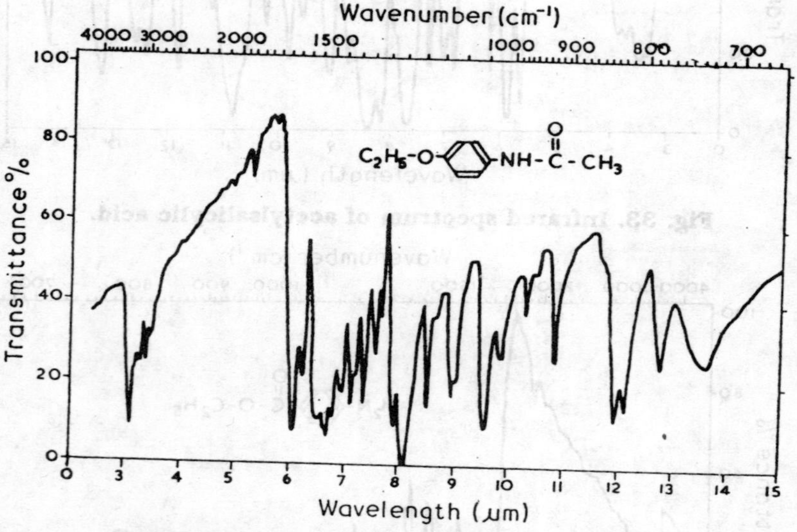


**Fig. 35. Infrared spectrum of phenacetin.**

Carbonyl absorption is most likely to account for the band at 1650 cm$^{-1}$, as double bond systems give rise to much less intense absorption. It is, however, rather low in frequency for an aldehyde, ketone or ester. The band at 1550 cm$^{-1}$ might be caused by NH deformation in an amide, by an amine or by a nitro group. When considered with the band at 1650 cm$^{-1}$, the most likely explanation is the presence of an amide group.

The absorption at 1250 cm$^{-1}$ is certainly in the C—O stretching region of an ester group, but aromatic ethers also absorb strongly in this region. Some confirmation for the group is given by the band at 1040 cm$^{-1}$.

Fig. 36-45 show the absorption of some common solvents.

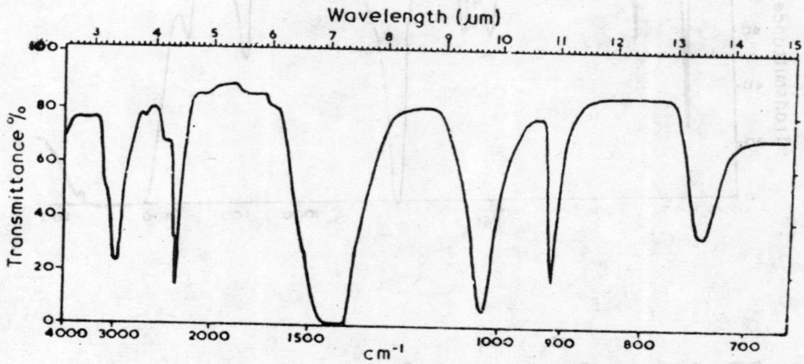


**Fig. 36. Infrared spectra of common solvents: Acetonitrile (0.1 mm cell).**

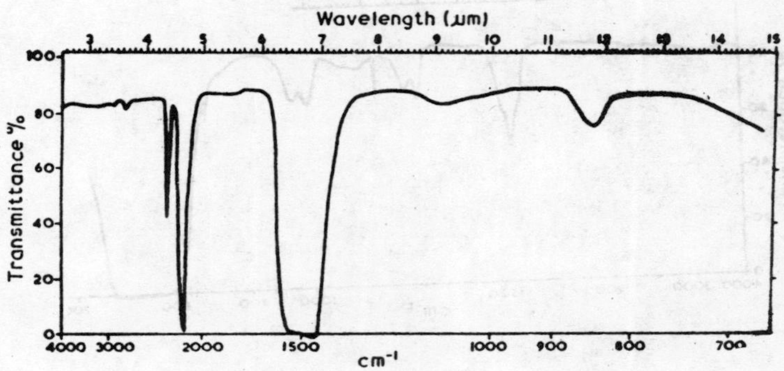


**Fig. 37. Infrared spectra of common solvents: Carbon disulphide (0.1 mm cell).**

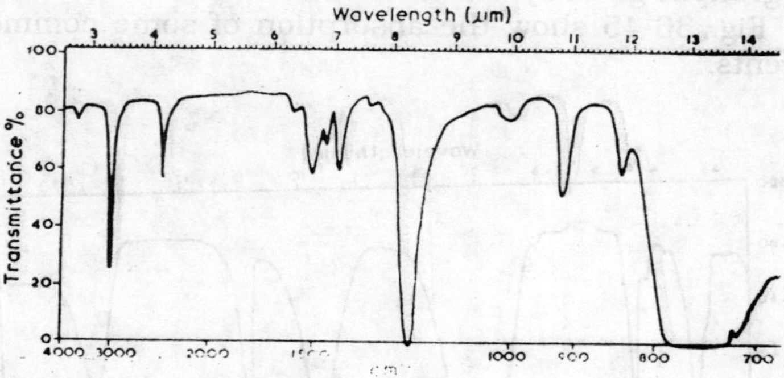


**Fig. 38. Infrared spectra of common solvents: Carbon tetrachloride (0.1 mm cell).**

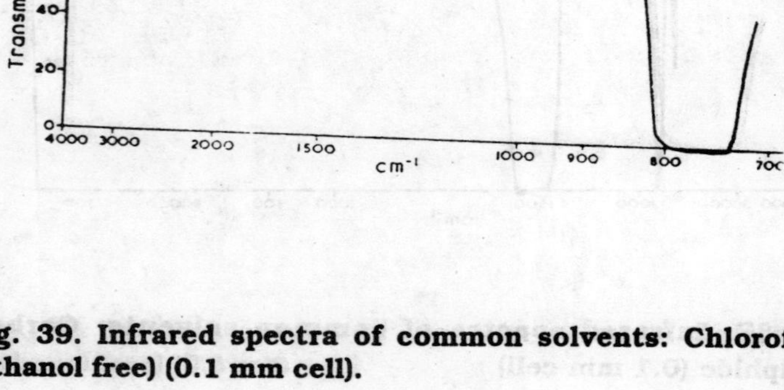


**Fig. 39. Infrared spectra of common solvents: Chloroform (ethanol free) (0.1 mm cell).**

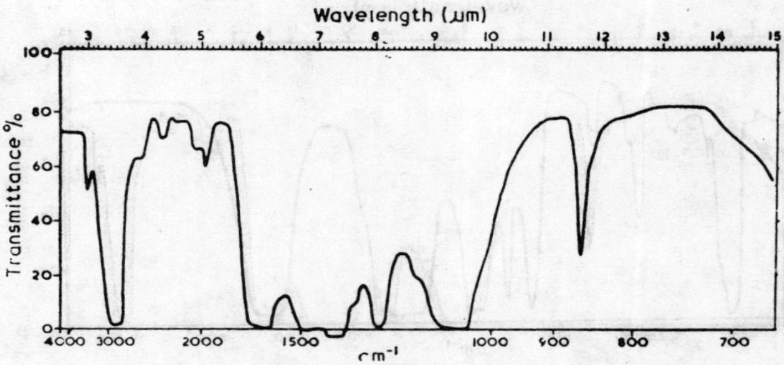


**Fig. 40. Fig. Infrared spectra of common solvents: Diemthylformamide (0.1 mm cell).**

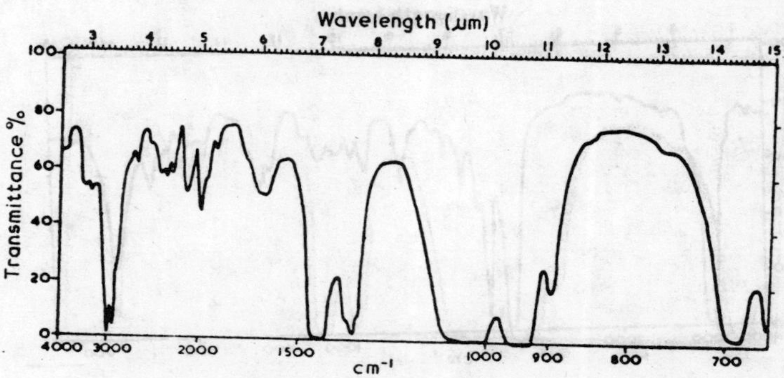


**Fig. 41. Infrared spectra of common solvents: Dimethyl sulphoxide (0.1 mm cell).**

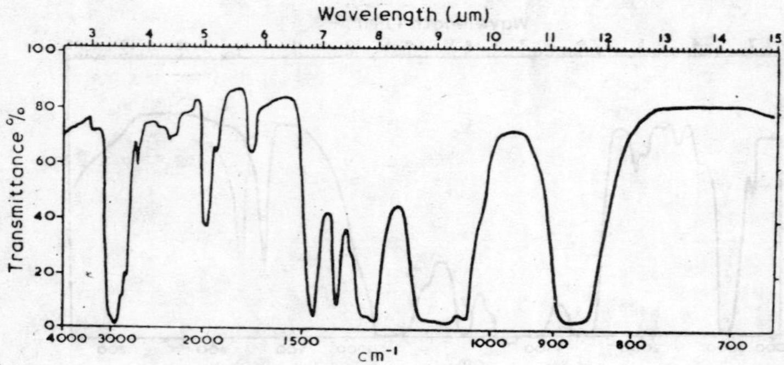


**Fig. 42. Infrared spectra of common solvents: Dioxan (0.1 mm cell).**

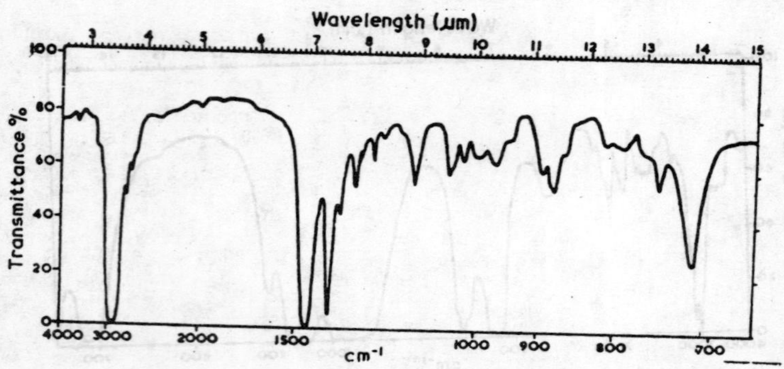


**Fig. 43. Infrared spectra of common solvents: n-Hexane (0.1 mm cell).**

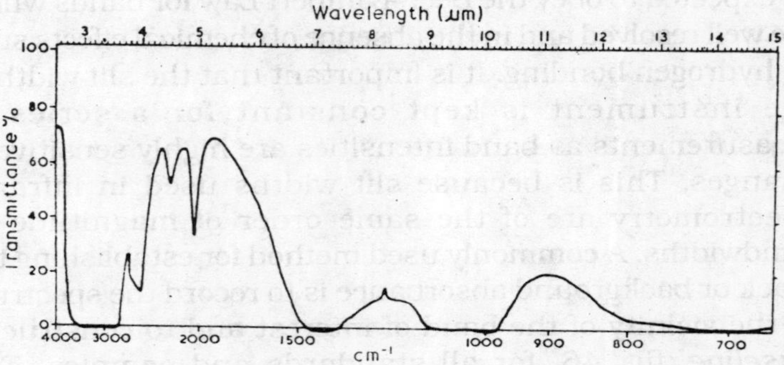


**Fig. 44. Infrared spectra of common solvents: Methanol (0.1 mm cell).**

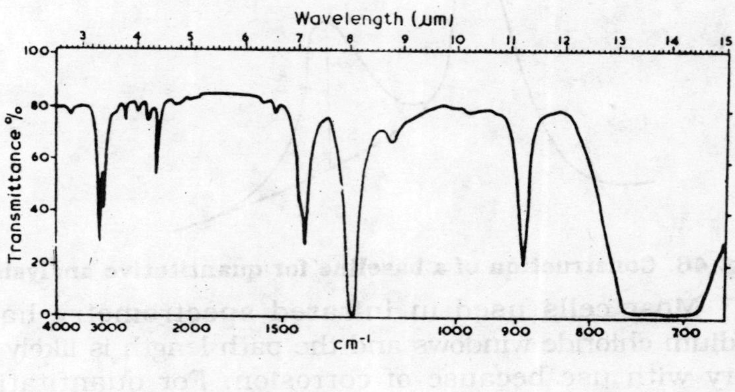


**Fig. 45. Infrared spectra of common solvents: Methylene chloride (0.1 mm cell).**

## Quantitative Analysis

Absorption in the infrared region of the spectrum can be expected to obey the Beer–Lambert Law for bands which are well resolved and in the absence of chemical effects such as hydrogen bonding. It is important that the slit width of the instrument is kept constant for a series of measurements as band intensities are highly sensitive to changes. This is because slit widths used in infrared spectrometry are of the same order of magnitude as bandwidths. A commonly used method for establishing the black or background absorbance is to record the spectrum in the vicinity of the band of interest and to construct a *baseline*, fig. 46, for all standards and samples. The baseline absorbance in measured at the point where a perpendicular line from the peak maximum intersects the baseline and this absorbance is subtracted from the peak absorbance in all cases.

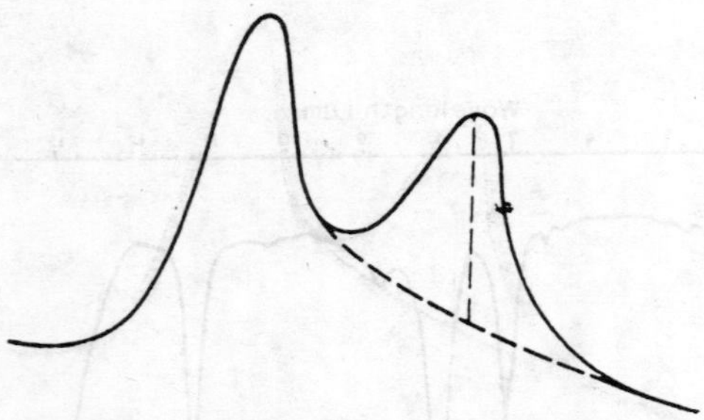

**Fig. 46. Construction of a baseline for quantitative analysis.**

Most cells used in infrared spectrometry have sodium chloride windows and the path length is likely to vary with use because of corrosion. For quantitative work, therefore, the same cell should be used for samples and standards. In general, quantitative analysis in the infrared region of the spectrum is not practised as widely as in the ultraviolet and visible regions, partly because of the additional care necessary to obtain reliable results and partly because the technique is generally considered to be less sensitive and less precise; a precision of 3 to 8% can be expected. However, the computer-aided analysis of solids by reflectance measurements, particularly in the near infrared region (NIR), is of growing importance. Although the OH, CH and NH overtones that occur in this region are weak (small molar absorptivities), radiation sources are more powerful, detectors more sensitive and the radiation penetrates more deeply into the sample surface than in the midinfrared region giving a more reliable measure of bulk composition. Quantitative calculations involve chemometrics in the form of computer controlled pattern recognition. Sets of equations generated from absorption measurements on many standards at multiple wavelengths throughout the NIR region are used in comparisons with corresponding data from the samples. Unfortunately, this approach is susceptible to the unexpected presence of a major impurity.

The principles underlying quantitative ultraviolet spectrophotometry apply also to infrared work. Moreover, infrared absorption spectra possess an advantage over those in the ultraviolet region in the greater number of bands present. It may often be possible to select a fairly strong band for each component in a mixture such that little or no interference occurs one with another. A calibration curve of extinction against concentration may be constructed or, if Beer's Law is shown to be obeyed, a direct comparison of sample and standard absorption may be used. Strictly, the integrated areas of the absorption bands should be compared but with sharp bands, peak heights can be used in calculations.

Infrared and ultraviolet spectrophotometry differ, however, in the concentrations used. Ten per cent solutions are common in infrared work and such a concentration is necessary because all solvents have some absorption in one part or another of the infrared region. This means that very short path lengths of 0.025 to 0.1 mm must be used in many assays. The high concentration of solute makes the accurate cancellation of solvent absorption very difficult but errors may be reduced by applying a base-line technique. The assumption is made that absorption due to solvent (or a second component) is constant or varies linearly with frequency over the region of the absorption band. The method of calculation is illustrated by reference to Fig. 47 *abc* is the recorded absorption of component A and *def* is the absorption caused by solvent and other components. Draw the line *agc* connecting the two minima *a* and *c* or between two suitable wavelengths on each side of the band. The point *g* is obtained by dropping a line perpendicular to the zero trasmittance line to meet *ac* to *g*. The extinction log $I_0I_T$ is calculated from the distances shown in Fig. 47.

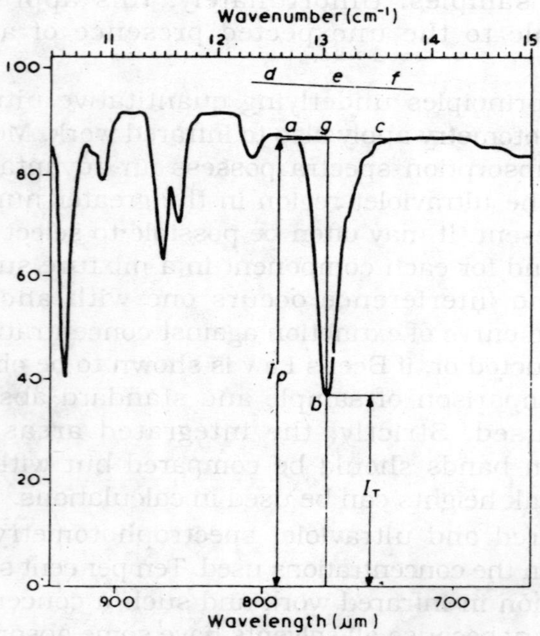


**Fig. 47. Infrared spectrum showing application of base-line technique.**

Even when bands do interfer, simultaneous equations may be set up as described for ultraviolet work. As many more components may be determined from an infrared spectrum than is possible with an ultraviolet absorption spectrum the work involved in solving a large number of simultaneous equations is considerable. Computers and matrix methods for solving simultaneous equations are, therefore, being used increasingly in analytical work. Infrared absorption assays also differ form ultraviolet absorption assays in the initial period of investigation necessary to obtain reproducible conditions. It may take some time to select the right solvent, cell thickness, slit width and absorption band.

For the determination of small quantities of substances, present either as impurity or as solute in a preparation, compensation fro absorption by the major component or solvent in the preparation may be achieved by introducing that component into the reference beam. A spectrum of the minor component only is thereby obtained. This method has been applied to the determination of β-picoline, 2,6-lutidine and 2-ethylpyridine in γ-picoline and also to solutions of testosterone propionate in arachis oil. Sufficient energy must be available for reliable results and the validity of Beer's Law for appropriate standard solutions will normally confirm this.

IR spectroscopy was generally considered to be able to provide only qualitative and semiquantitative analyses of common samples, especially when the data were acquired using the conventional dispersive instruments. However, the development of reliable FTIR instrumentation and strong computerized data-processing capabilities have greatly improved the performance of quantitative IR work. Thus, modern infrared spectroscopy has gained acceptance as a reliable tool for quantitative analysis.

The basis for quantitative analysis of absorption spectrometry is the Bouguer–Beer–Lambert law, commonly called Beer's law. For a single compound in a homogeneous medium, the absorbance at any frequency is expressed as

$$A = abc \quad (1)$$

where $A$ is the measured sample absorbance at the given frequency, $a$ is the molecular absorptivity at the frequency, $b$ is the path length of source beam in the sample, and $c$ is the concentration of the sample. This law basically states that the intensities of absorption bands are linearly proportional to the concentration of each component in a homogeneous mixture or solution.

Deviations from Beer's law occur more often in infrared spectroscopy than in UV/visible spectroscopy. These deviations stem from both instrumental and sample effects. Instrumental effects include insufficient resolution and stray radiation. Resolution is closely related to the slit width in dispersive IR instruments or the optical path difference between two beams in the interferometer of FTIR spectrometers. Stray light levels in FT instruments are usually negligible. Sample effects include chemical reactions and molecular interactions such as hydrogen bonding. The Beer's law deviations result in a nonlinear relationship for plots of absorbance ($A$) against concentration ($c$). It is therefore a good practice to obtain calibration curves that are determined empirically from known standards.

Instead of the transmittance scale, absorbance is generally used in quantitative analysis. Absorbance ($A$) is defined as the negative logarithm of the transmittance ($T$). According to Beer's law, a linear relationship exists only between the sample concentration and absorbance, not between the sample concentration and transmittance. The linearity of Beer's law plots usually holds better when the absorbance is limited to less than 0.7 absorbance units, although in some case good linearity has been achieved over more than 2 absorbance units. A number of quantification parameters, which include peak height, peak area, and derivatives, can be used in quantitative analysis.

The integration limits for peak area determinations should be carefully chosen to ensure maximum accuracy.

In multicomponent quantitative analysis, the determination of the composition of mixtures involves the use of software packages. These analyses usually assume that Beer's law is additive for a mixture of compounds at a specified frequency. For a simple two-component mixture, the total absorbance, $A_T$, of the mixture at a given frequency is the sum of the absorbance of two component compounds, $x$ and $y$, at the specified frequency :

$$A_T = A_x + A_y = a_x bc_x + a_y bc_y \tag{2}$$

It is necessary to determine $a_x$ and $a_y$ from absorption measurements of mixtures containing known amounts of compounds $x$ and $y$ at two different frequencies, $n$ and $m$. Using these values, $a_{x,n}$, $a_{x,m}$, $a_{y,n}$, and $a_{y,m}$, it is possible to use two absorbance measurements from the mixture of unknown composition to determine the concentrations of compound $x$ and $y$, $c_x$ and $c_y$.

$$A_{T,n} = A_{x,n} + A_{y,n} = a_{x,n} bc_x + a_{y,n} bc_y \tag{3}$$

$$A_{T,m} = A_{x,m} + A_{y,m} = a_{x,m} bc_x + a_{y,m} bc_y \tag{4}$$

Using matrix algebra it is possible to extend this technique to mixtures containing more than two components. The absorbance of mixture of $n$ independently absorbing components at a particular frequency v may be expressed in the following equation:

$$A_v = a_1 bc_1 + a_2 bc_2 + \ldots + a_n bc_n \tag{5}$$

where $A_v$ = total absorbance of the sample at the frequency, v, $a_j$ = absorptivity of component $j$ at the frequency v ($j$ = 1,2, ... $n$), $c_j$ = concentration of component $j$, and $b$ = sample path length.

Software packages containing matrix methods available with computerized spectrometers simplify the operations associated with multicomponent analysis. If deviations of Beer's law occur, but the law of addivity still holds, sophisticated correlation or statistical evaluation software programs such as least-squares regression, partial least-squares regression, and principal component regression analysis facilitate satisfactory curve-fitting and data-processing tasks.

The broad absorption bands, larger values of absorptivity and sample path length, higher intensity sources, and more sensitive detectors make the ultraviolet, visible, and near IR regions better suited for quantitative determinations than the mid IR and far IR regions. However, coupling of the advancement of computerized FTIR instrumentation and molecules attention to detail can make FTIR a viable option for reliable quantitative analysis.

## Application of Infrared Spectroscopy

The unique appearance of an infrared spectrum has resulted in the extensive use of infrared spectrometry to characterize such materials as natural products, polymers, detergents, lubricants, fats and resins. It is of particular value to the petroleum and polymer industries, to drug manufacturers and to producers of organic chemicals. Quatitative applications include the quality control of additives in fuel and lubricant blends and to assess the extent of chemical changes in various products due to ageing and use. The food and grain industries use quantitative reflectance measurements in the NIR for the determination of water, fats, proteins and sugars. Non-dispersive infrared analysers are used to monitor gas streams in industrial processes and atmospheric pollution. The instruments used are generally portable and robust, consisting only of a radiation source, reference and sample cells and a detector filled with the gas which is to be monitored.

### 1. Analysis of Petroleum Hydrocarbons, Oil, and Grease Contents by EPA Methods

The Environmental Protection Agency (EPA) has established Methods 413.2 and 418.1 for the measurement of fluorocarbon-113 extractable matter from surface and saline waters and industrial and domestic wastes. These methods provide semiquantiative determination of petroleum hydrocarbons, oil, and grease by comparison of the infrared absorption of the sample extract with standards.

Petroleum hydrocarbons, oil, and grease include biodegradable animal greases and vegetable oils along with the relative nonbiodegradable mineral oils. They all contain carbon–hydrogen bonds, thus giving rise to C–H stretching absorption in the 3100 to 2700 $cm^{-1}$ region of the IR spectrum. Fluorocarbon-113 (1,1,2-trichloro-1,2,2-trifluoroethane) is one of the chlorofluoro-carbons commonly called freons. It contains no C—H bonds, and thus does not absorb IR radiation in the aforementioned 3100 $cm^{-1}$ region. The quantity of hydrocarbons, oil, and grease in freon extracts can be estimated by measuring the intensity of C–H absorption band at 2930 $cm^{-1}$.

The sample is acidified to a low pH (less than 2) and extracted with fluorocarbon-113. Interference is usually removed with silica adsorbent. Depending on the sample concentration, cells of pathlength from 10 to 100 mm can be used to acquire the normal transmission IR spectrum. The concentration of hydrocarbon, oil, and grease in the extract is determined by comparing the absorbance against the calibration plot prepared from standard calibration mixtures. Fig. 48 shows the different FTIR profiles of three calibration standards. The contributions from the solvent and cell are eliminated by subtracting the reference spectrum of freon from the sample spectrum.

In the standard EPA methods, peak height at a single frequency, 2930 $cm^{-1}$, is used as the basis for quantification. In the author's laboratory, peak area integration from 3150 to 2700 $cm^{-1}$ is used to quantify the contents of hydrocarbons, oil, and grease. The modified methods have been found to provide significantly improved results in quantitative analysis. Aromatic hydrocarbons have relatively lower absorption intensity in this C–H stretching region, thus giving lower response factors when compared to the IR absorption of oil and grease standards. Using an FTIR instrument, oil and grease at low parts-per-million levels can be readily determined.

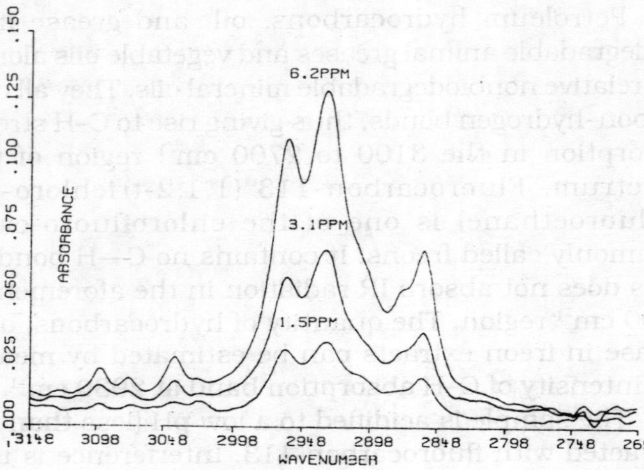


**Fig. 48. FTIR spectra of three calibration standards, at 1.5, 3.1, and 6.2 ppm, for oil and grease analysis.**

## 2. Quantitative Analysis of Multicomponent Mixtures of Sulfur Oxygen Anions by Attenuate Total Reflectance Spectroscopy

Characterization of complex mixtures of sulfur oxygen anions is encountered in studies such as investigating the decomposition of the dithionite anion in acidic aqueous solution. Many techniques such as conventional IR, UV/visible, Raman spectroscopy, and titrimetric and electrochemical analysis all have drawbacks that limit their effectiveness in this challenging analysis. The multicomponent analysis by FTIR attenuated total reflectance (ATR) spectroscopy successfully provides the accurate quantification of multiple sulfur–oxygen anion concentrations in aqueous solution (5).

Sulfur–oxygen compounds have relatively intense S-O stretching absorption bands in the 1350 to 750 $cm^{-1}$ region of the IR spectrum. FTIR/ATR spectroscopy is well suited for quantitative determination of sulfur oxygen anions in strong IR-absorbing aqueous medium. ATR not only uses water-resistant cell material, but also has a very short and reproducible effective path length that goes beyond the sample/crystal interface and into the sample medium. A micro CIRCLE cell from Spectra-Tech incorporating a ZnSe crystal is used in the study. Its basic optics is illustrated in Fig. 49. The representative FTIR/ATR spectra of sulfur oxide and nitrate anions are shown in Fig. 50.

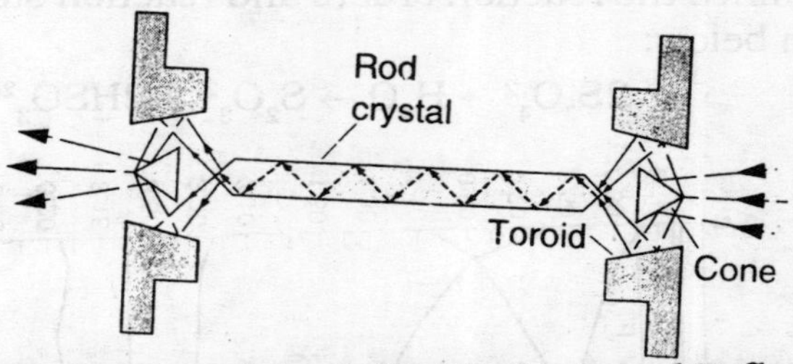


**Fig. 49. Optic diagram of the Cylindrical Internal Reflectance Cell for Liquid Evaluation (CIRCLE).**

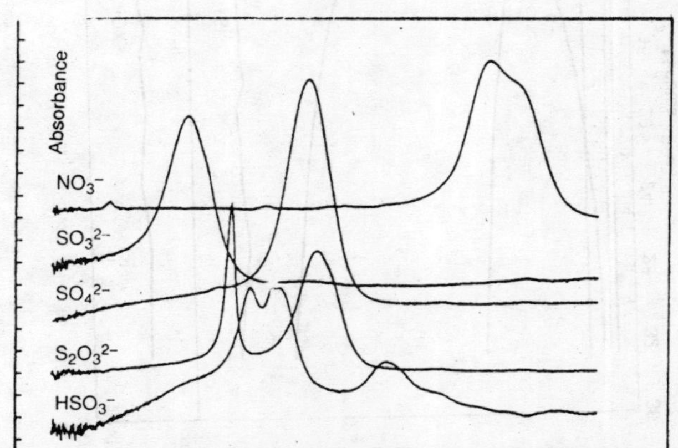


**Fig. 50. Representation of FTIR/ATR spectra of nitrate and four sulfur oxide anions.**

Aqueous decomposition of sodium dithionite under anaerobic conditions is investigated. Systematic beaseline error is characterized and taken into account. Computerized data processing involving least-squares regression is used in the multivariate analysis. Fig. 51 illustrates the simultaneous measurements of concentrations of seven anions and continuous monitoring of total sulfur and average oxidation states over a 30-min reaction period. This analytical technique successfully determines the reaction orders and reaction stoichiometry shown below:

$$2S_2O_4^{2-} + H_2O \rightarrow S_2O_3^{2-} + 2HSO_3^{2-} \qquad (1)$$

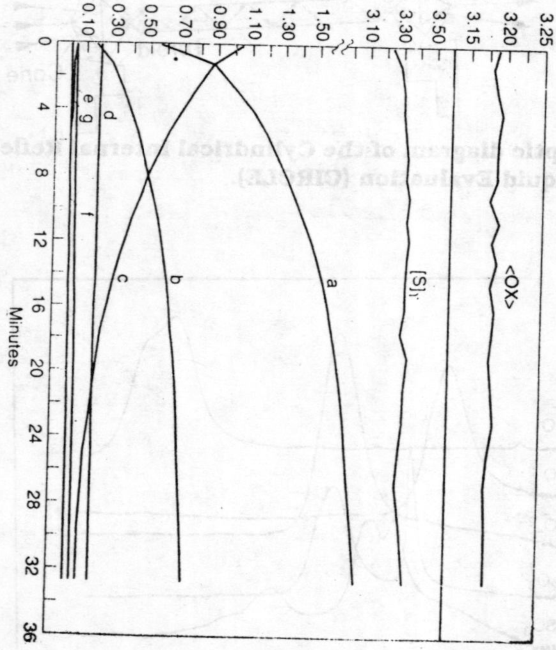

**Fig. 51. Multicomponent analysis of sulfur oxygen anions over a 30-min reaction period of the anaerobic aqueous decomposition of sodium dithionite. (a) [$HSO_3^-$], (b) [$S_2O_3^{2-}$], (c) [$S_2O_4^{2-}$], (d) [$S_3O_6^{2-}$], (e) [$SO_4^{2-}$], (f) [$S_2O_5^{2-}$], (g) [$SO_3^{2-}$], (OX) = average oxidation state, and [S]$_T$ = total sulfur.**

## 3. Characterization of Heterogeneous Catalysts by Diffuse Reflectance Spectroscopy

Vibrational spectroscopy has been long established as one of the preferred techniques in obtaining important information on the nature of molecules that are attached to the catalyst surfaces. Diffuse reflectance (DR) spectroscopy has been shown to provide additional advantages over the conventional transmission methods. Detailed characterization of an olefin polymerization catalyst is described here to demonstrate one of the practical applications (6).

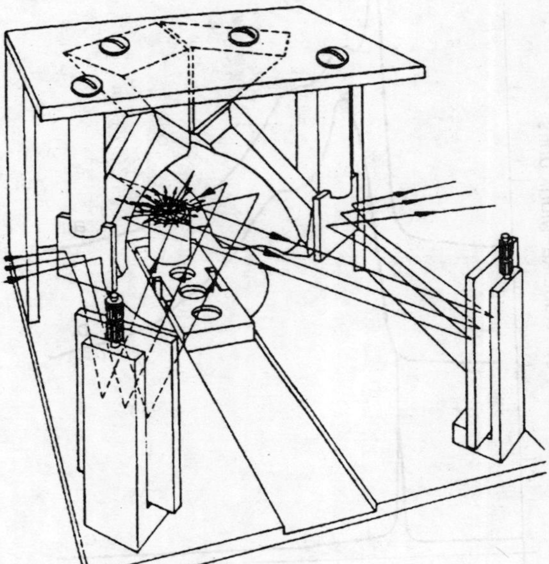

**Fig. 52. Optical diagram of a diffuse reflectance accessory.**

Heterogenous catalysts often consists of active molecules in a distribution of valence states supported on high-surface-area oxides, such as silica and alumina. To better understand Zigler–Natta catalysts (important catalysts for olefin polymerization processes), the reactions of attaching titanium chloride ($TiCl_4$) to modified silica surfaces are studied. A diffuse reflectance accessory from Harrick

Scientific (Fig. 52) and a controlled-atmosphere, high-temperature cell are used in the experiments.

Catalyst samples are diluted by making a 10% w/w dispersion in dry KCl power. Diffuse reflectance spectra of 200°C pretreated silica gel before and after reaction with hexamethyldisilazane (HMDS) and 600°C pretreated silica gels are shown in Fig. 53.

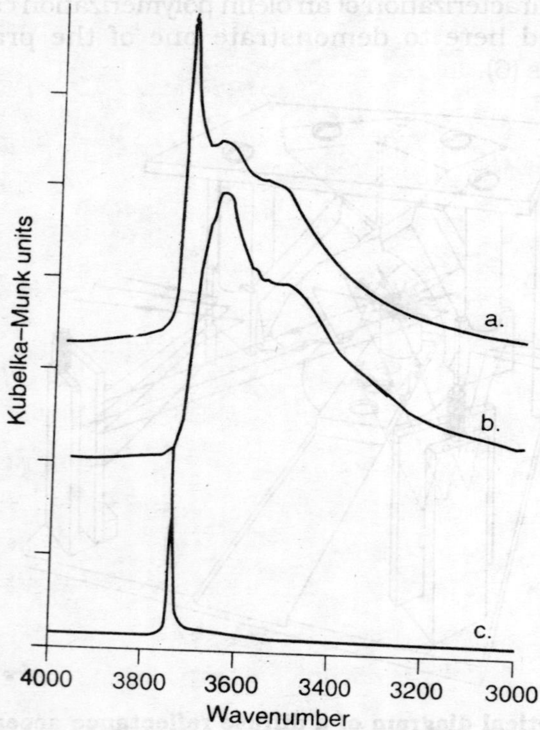


**Fig. 53. FTIR diffuse reflectance spectra of silica gel samples. (a) 200°C pretreated silica, (b) HMDS pretreated silica, and (c) 600°C pretreated silica.**

The chemical reactions on modified silica gel surfaces are illustrated in Fig. 54. The 200°C pretreated silica gel exhibits three surface hydroxyl absorption bands at 3740, 3660, and 3540 $cm^{-1}$, which arise from relatively free, non-hydrogen-bonded, and hydrogen-bonded silanols (Si-OH), respectively. After surface modification with HMDS treatment, the 3740 $cm^{-1}$ peak disappears while the broader bands from hydrogen-bonded silanols remain relatively unperturbed. Baking of the silica gel to 600°C results in the disappearance of peaks at 3660 and 3540 $cm^{-1}$, but does not significantly affect the non-hydrogen-bonded silanols' absorption at 3740 $cm^{-1}$. Diffuse reflectance spectra of 200°C pretreated silica gel before and after $TiCl_4$ reaction, along with the spectrum resulting from spectral subtraction are shown in Fig.55. The IR absorption bands at 990 and 920 $cm^{-1}$ in the difference spectrum are assigned to the Si-O stretchings of Si-O-$TiCl_3$ and $(Si\text{-}O)_2$-$TiCl_2$, respectively. Similar analyses support the other reactions illustrated in Fig. 54.

## 4. Qualitative Analysis of Multilayered Polymeric Films using FTIR Mircospectroscopy

IR analysis can be used to determine the identities of polymer materials in a multilayered film. Using FTIR microspectroscopy, various layers in the polymeric film can be quickly characterized. The qualitative analysis of a three-layer, 20-µm-thick film is described below to demonstrate such an application.

The layered thin film is cut as a 2×20-mm sheet. Cross-sections of the film are obtained using a fiber microtome. The individual section is transferred to a NaCL window on a slide positioned on the microscope state. FTIR spectra are recorded in the transmission mode.

This particular sample has a three-layer composition of Primacor (8 µm), nylon (10 µm), and Primacro (8 µm). Primacor is a copolymer of ethylene and acrylic acid. Nylon is a polyamide polymer. Using an FTIR microscope, infrared transmittance spectra of the multilayered film are obtained. An IR spectrum of pure Primacor can be obtained on the exposed outside layer. The IR spectrum of the center layer exhibits the contributions from both nylon and Primacor. This probably results from spatial contamination, which occurs when a specific layer of 8 to 10 µm or thinner is not masked properly for IR spectrum acquisition due to the poor contrast between the layers or the limitations of the aperture sizes. Functional, group mapping can be performed to enhance the spatial resolution. This is accomplished by driving microscope stages across the film cross-section while mapping the concentration of functional groups that represent characteristics of each component. This type of infrared imaging technique can be more effectively carried out with computer-controlled, two-dimensional motorized stage and provides a systematic, nondestructive evaluation of sample composition on a microscopic scale.

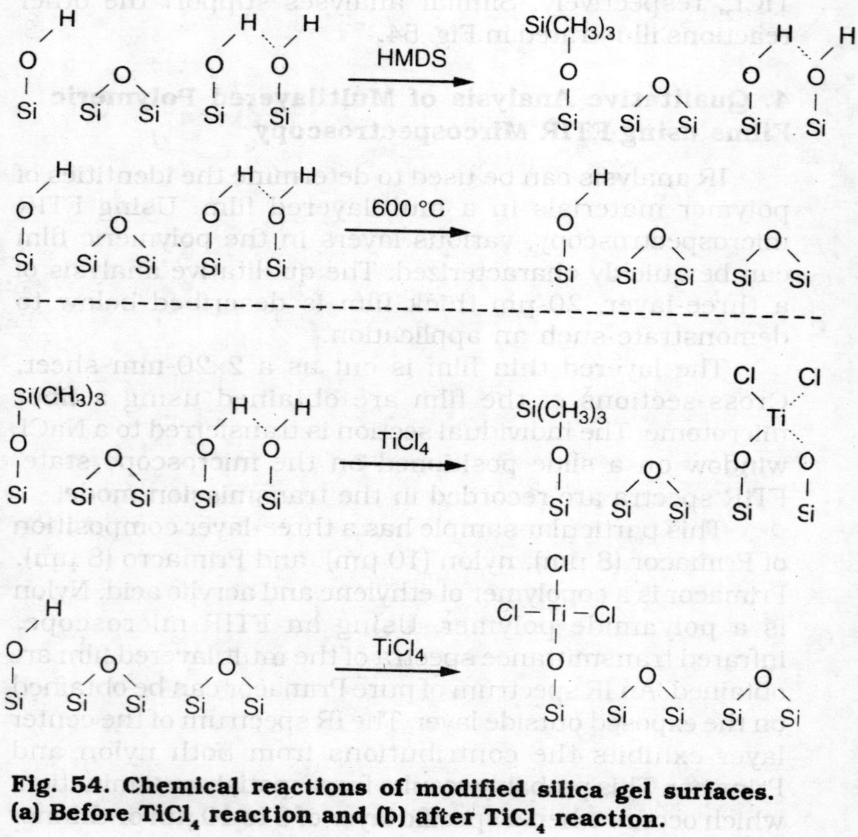


**Fig. 54. Chemical reactions of modified silica gel surfaces. (a) Before $TiCl_4$ reaction and (b) after $TiCl_4$ reaction.**

## Experiments

### Experiment 1. Determination of the absorption spectrum of a liquid–cyclohexanol

*Method.* Check the instrument for 100% and 0% transmission and select the slit width programme (minimum slit width). Place a drop of the liquid on a sodium chloride plate of a demountable cell and cover with another sodium chloride plate thus forming a capillary film of liquid. Insert the cell into the sample beam. If the film is too thick, carefully tighten the screws holding the cell until a transmission of about 5% is obtained for the strongest absorption band. Do not overtighten or the plates may crack.

After use, dismantle the cell, rinse the plates with chloroform and allow to dry under a lamp to avoid condensation of water vapour as the chloroform evaporates.

The thickness of the film can be varied by means of lead spacers but the method adopted above proves satisfactory in most cases. The liquid chosen here illustrates, in conjunction with the results of Experiment 5, the effect of hydrogen bonding on the absorption spectrum. Phenols, alcohols and carboxylic acids show this effect, and for further discussion.

### Experiment 2. Determination of the absorption spectrum of a solid using the mull technique

*Method.* Powder thoroughly about 15–20 mg of sample in a small agate mortar and add 2 drops of purified liquid paraffin. Continue the trituration until a smooth paste is obtained and transfer it to one of the sodium, chloride plates of a demountable cell. Assemble the cell and continue as described for a liquid in Experiment 2.

It is essential to reduce the particle size of the sample to below 3 μm otherwise scattering of radiation will give rise to a poor absorption spectrum. With care, as little as 3 mg of material may be used with a small amount of liquid paraffin. In this method unless the material dissolves in the paraffin, the spectrum obtained is that of the *solid* sample. Crystal forces and hydrogen bonding may, therefore, influence the trace obtained. Strong bands due to paraffin itself appear at 2920–2850 and 1460–1380 $cm^{-1}$.

**Experiment 3. To determine the path length of a sealed cell**

*Method.* Record the spectrum of the *empty* cell versus air and note the wavelengths between which lie an integral number of fringes. Calculate the thickness of the cell by means of the equations:

(i) $$t \text{ (in cm)} = \frac{n}{2}\left(\frac{1}{\bar{v}_1 - \bar{v}_2}\right)$$

where n = numbers of fringes and $\bar{v}_1, \bar{v}_2$ = wave numbers selected.

or

(ii) $$t \text{ (in } \mu\text{m)} = \frac{n}{2}\left(\frac{\lambda_1 \times \lambda_2}{\lambda_1 - \lambda_2}\right)$$

where $\lambda_1$, $\lambda_2$ = wavelengths selected (μm).

The 'spectrum' obtained in this experiment is that of a typical interference fringe pattern and may be obtained for cells up to about 0.5 mm in path length. To illustrate the calculation refer to the fringe pattern in Fig.55.

$$t = \frac{6}{2}\left(\frac{5 \times 10.1}{10.1 - 5}\right)$$

$$= 29.7\ \mu\text{m}$$

$$= 0.030\ \text{mm}$$

The appearance of the pattern itself often indicates the quality of the cell: a poor fringe pattern is given by cells in which the plates are not exactly parallel or have become worn after use. Compare the patterns for *a* and *b* in Fig. 56.

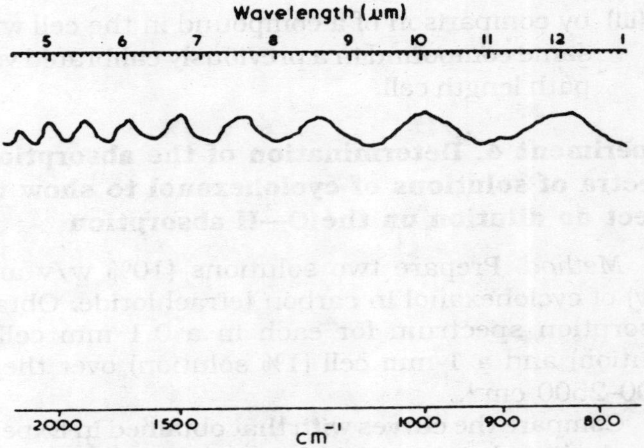


**Fig. 55. Interference fringe pattern.**

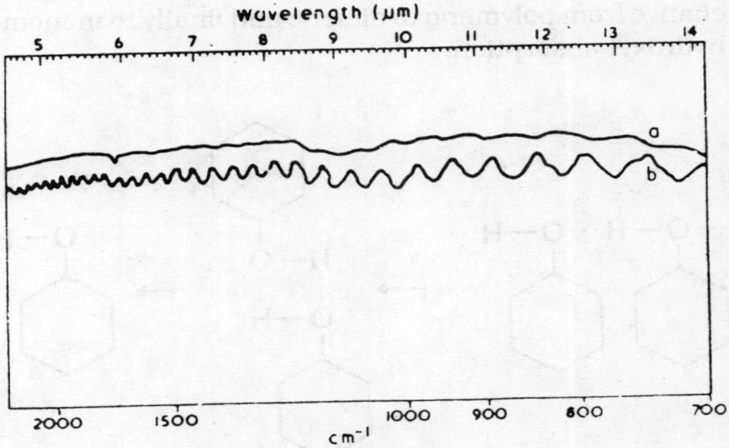


**Fig. 56. Interference fringe pattern.**

For cells longer than about 0.5 mm the path length maybe determined in several ways:

(i) by examination of a substance whose extinction coefficient is accurately known

(ii) by use of a travelling microscope

(iii) by comparison of a compound in the cell with the same compound in a previously calibrated variable path length cell.

### Experiment 4. Determination of the absorption spectra of solutions of cyclohexanol to show the effect do dilution on the O—H absorption

*Method.* Prepare two solutions (10% w/v and 1% w/v) of cyclohexanol in carbon tetrachloride. Obtain the absorption spectrum for each in a 0.1 mm cell (10% solution) and a 1 mm cell (1% solution) over the range 4000–2500 cm$^{-1}$.

Compare the curves with that obtained in Experiment 2. In the liquid state, a broad absorption band is recorded and this is retained in the strong (10%) solution. Note how this is reduced in the 1% solution; note also the increase in absorption at shorter wavelengths. This effect is due to change from polymeric to dimeric and finally to monomeric hydroxyl absorption.

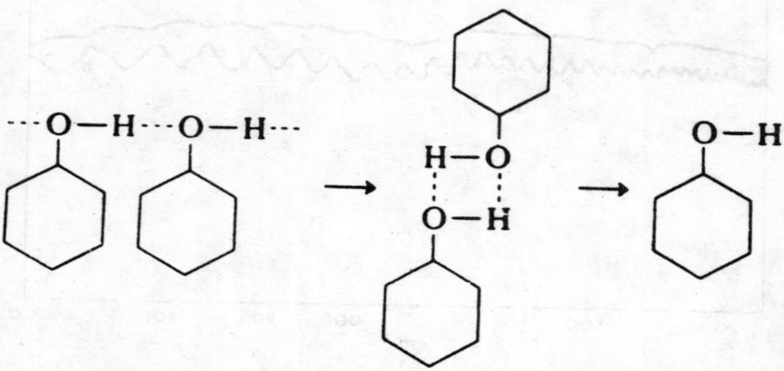


max ≈ 3300 cm$^{-1}$ (broad)

max ≈ 3610 cm$^{-1}$ (sharp)

The formation of hydrogen bonds will tend to weaken the O—H bond, hence *k*, the force constant, will decrease, and the observed vibrational frequency will be less than that for the monomeric form. Carboxylic acids, however, remain dimeric even on dilution. If the hydrogen bond is intramolecular, no change in the O—H absorption will occur on dilution.

Solution spectra are useful in that difficulties inherent in solid state spectra are absent. However, solvent effects may occur as a shift of the absorption frequency of a substance when one solvent is substituted for another. All solvents absorb in some part of the infrared region and, therefore, several solvents are necessary to cover the whole spectrum. In this experiment carbon tetrachloride transmits most of the incident light and no reference cell is required.

The concentration needed for solutions will vary according to the substance and cell thickness, but a useful concentration for preliminary investigation is 10 per cent w/v in a 0.1 mm cell.

Very thin films of aqueous solutions can be examined in cells of barium fluoride, Irtran-2 or silver chloride but these solutions and, indeed, almost any kind of sample can be handled by a technique called attenuated total reflectance, which is based upon total internal reflectance. The efficiency of the interface between two media as a mirror is dependent on the difference in refractive indices of the media. As the refractive index of a material changes rapidly in the region of an absorption band these changes result in a decreased amount of energy reflected at the interface. The resulting spectra are, strictly, plots of reflected energy against wavelength and appear very similar to conventional absorption bands; they provide essentially the same information.

The technique requires an accessory for the infrared spectrophotometer and the basic design for the unit handling micro-samples is shown in Fig. 57. Sample are held in contact with the flat face of the crystal hemicylinder and different crystals are required to provide a satisfactory range of reflective indices to cover all samples. Typical crystal materials are KRS-5 germanium and silver chloride;

giving a high, medium and low refractive index respectively. The rotating mirrors are necessary to allow various angles of incidence on the crystal so that optimum attenuated total reflectance spectra are obtained.

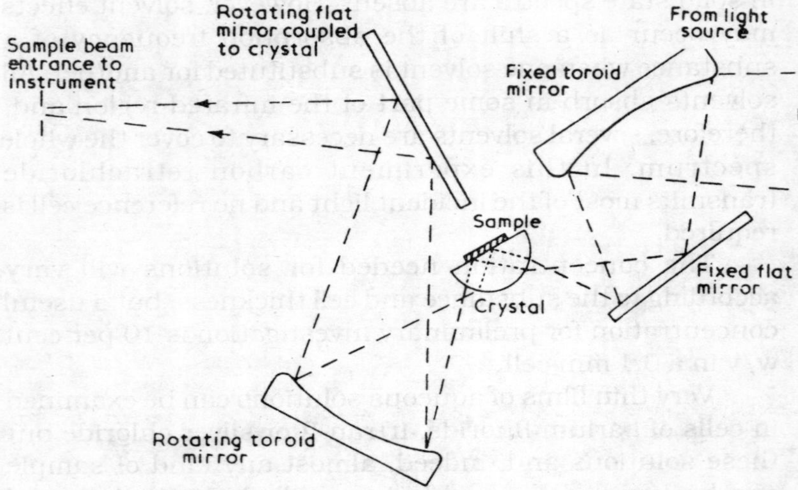


**Fig. 57. Unit for handling micro samples.**

## Cognate Experiments

*Benzoic acid*

*Catechol*

*o,o' and p,p'-Dihydroxydiphenyl*

## Experiment 5. Determination of the absorption curve of a solid using the potassium chloride disc technique

*Method.* Finely powder the sample (1 mg) and mix with dry finely powdered potassium chloride (200 mg). Triturate the mixture thoroughly for about 2 min and transfer to a suitable die and press. Attach the die to a vacuum pump, allow to stand for 1 min under vacuum and compress the powder at 10–12 tons/sq. inch for 2 min. Release the pressure and vacuum, and extract the disc for insertion in the sample beam of the spectrophotometer.

For potassium chloride alone the disc should be completely transparent, but it may become translucent when the sample is present. The disc is about 13 mm diameter and the above quantities will, therefore, give a weight of sample between 10 and 15 μg/mm$^2$.

Attention has already been drawn to the care necessary when using this technique for comparison purposes. It has the advantage, however, of giving spectra free from solvent peaks, and is, therefore, extremely useful as a routine method.

The disc method has not entirely superseded the mull technique for the spectra of solid samples because differences have been observed in the results obtained by using the two techniques; the differences are related to the treatment the sample received during the preparation of the disc. In finely ground material, for example, interaction may occur between vibration so the sample and the potassium halide lattice. For a discussion of the infrared analysis of solid substances.

The method is also used in the examination of the very small quantities of material eluted from columns in gas chromatography. The fraction is trapped in about 300 mg of potassium chloride in the form of a short column placed immediately after the detector. The solid is powdered, pressed into a disc in the normal way and the absorption spectrum of the trapped fraction is determined.

### Experiment 6. Determination of the amount of phenobarbitone in Phenobarbitone Tablets

*Method.* Prepare a series of solutions of phenobarbitone in chloroform containing 20, 40, 60, 80, and 100 mg in each 10 ml. Transfer each solution in turn to a 0.1 mm cell and measure the absorption curve over the region 4.4–6.5 μm. Calculate the absorbance using the base-line technique for each solution at the peak absorption (about 1730 $cm^{-1}$) and plot a graph of absorbance against concentration in the normal manner.

Weight and powder 20 tablets. Extract with chloroform an accurately weighed quantity of powdered tablets equivalent to about 80 mg of phenobarbitone. Evaporate the extract to dryness and dissolve in sufficient chloroform to produce 10 ml of solution. Determine the absorption curve as described above and calculate the absorbance at the selected peak.

Read off the number of mg of phenobarbitone from the graph. Hence calculate the number mg of phenobarbitone in a tablet of average weight.

Careful standardisation of instrumental conditions is essential in quantitative analysis. Of these the 100% and 0% transition and the slit width settings are very important. If the calibration curve obeys Beer's Law, a direct comparison of the absorbance of test and standard solutions at the peak absorption may be used. This direct comparison was adopted in the United States Pharmacopeia XVI for the assay of Acetazolamide and Diethyltoluamide.

### Experiment 7. Determination of the amount of acetylsalicylic acid and phenacetin in Acetylsalicylic Acid and Phenacetin Tablets

*Method. (i) Calibration curves.* Carefully check the 100% and 0% transmission over the 8–10 μm range and adjust, if necessary. Prepare the following solutions using chloroform as solvent :

| | |
|---|---|
| Acetylsalicylic acid | 1,2,3 and 4% w/v |
| Phenacetin | 1,1.5,2 and 2.5% w/v |

Record in a 0.1 mm cell the absorption curve of each solution against a blank of chloroform in the reference beam. With chloroform alone the 100% line is obtained. Calculate absorbances for each solution at two wavelengths— about 8.95 μm at which phenacetin absorbs and at about 9.22 μm at which acetylsalicylic acid absorbs. Plot the calibration curves as shown in Fig. 5 which were obtained in an actual experiment.

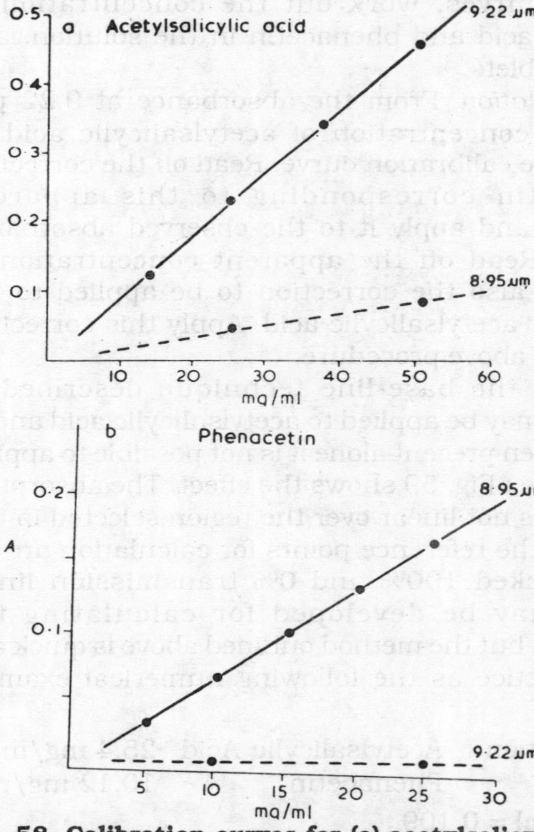


**Fig. 58. Calibration curves for (a) acetylsalicylic acid and (b) phenacetin**

(ii) *Tablets.* Weigh and powder 20 tablets. Transfer an accurately weighed quantity of the powder equivalent to about 0.3 g of acetylsalicylic acid to a sintered-glass crucible and wash with small portions of chloroform until the acetylsalicylic acid and phenacetin are completely extracted. Collect the washings and evaporate to dryness. Dissolve the residue in chlorform and transfer quantitatively to a 10 ml flask. Make up to volume with chloroform, mix, and record the absorption curve over the 8–10 μm region using a blank of chloroform. Calculate the absorbances at about 8.95 and 9.22 μm and, by reference to the calibrations curves, work out the concentration of acetylsalicylic acid and phenacetin in the solution, and hence in the tablets.

(iii) *Calculation.* From the absorbance at 9.22 μm calculate the concentration of acetylsalicylic acid by reference to the calibration curve. Read off the correction for phenacetin corresponding to this apparent concentration and apply it to the observed absorbance at 8.95 μm. Read off the apparent concentration of phenacetin of also the correction to be applied to the absorbance for acetylsalicylic acid. Apply this correction and repeat the above procedure.

Although the base-line technique described in Experiment 7 may be applied to acetylsalicylic acid and to phenacetin when present alone it is not possible to apply it in this instance : Fig. 59 shows the effect. The absorption curve of each is not linear over the region selected in this assay so that the reference points for calculation are the carefully checked 100% and 0% transmission lines. Equations may be developed for calculating the concentrations but the method outlined above is quick and simple in practice as the following numerical example shows.

| Test Solution : | Acetylsalicylic Acid | 25.4 mg/ml |
|---|---|---|
| | Phenacetin | 10.12 mg/ml |

$A$(8.95 μm) = 0.109
$A$(9.22 μm) = 0.236

First calculation using $E$(9.22 μm) gave acetylsalicyclic acid as 28.0 mg/ml. The correction to the absorbance at 8.95 μm was 0.45 (from graph)

∴ Phenacetin = 9.4 mg/ml.

Second calculation. The correction to the absorbance for acetylsalicylic acid because of the phenacetin was 0.0175.

∴ Acetylsalicylic acid = 25.4 mg/ml.

Using this concentration the correction to be applied to the absorbance for phenacetin was 0.040.

∴ Phenacetin 10.1 mg/ml.

Third calculation.

Correction for acetylsalicylic acid= 0.018

∴ Acetylsalicylic acid = 25.4 mg/ml.

Correction for phenacetin = 0.40

∴ Phenacetin = 10.1 mg/ml.

The results of the second and third calculations are identical and, therefore, no further calculations need be performed.

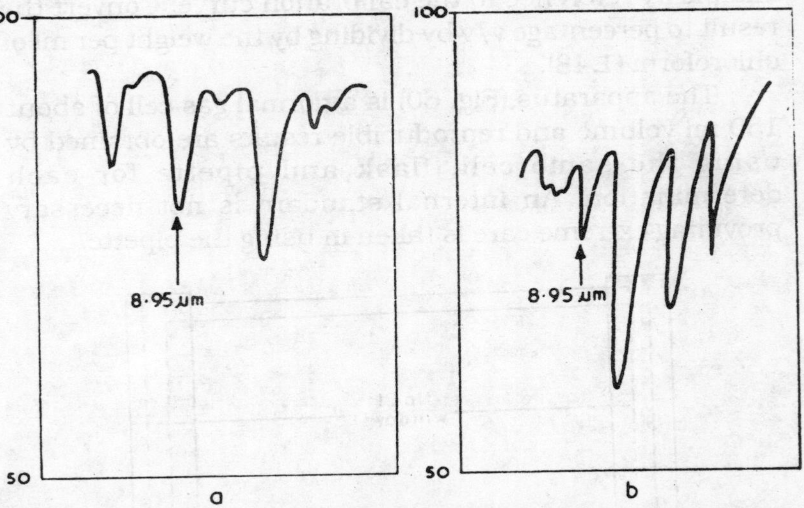


**Fig. 59. Absorption curve of (a) phenacetin (10.1 mg/ml) in chloroform, and (b) phenacetin (10.1 mg/ml) in chloroform containing acetylsalicylic acid (25.4 mg/ml).**

## Experiment 8. To determine the percentage v/v of chloroform in Chloroform Spirit

*Method.* Prepare a series of solutions of ethanol-free chloroform in ethylene glycol to contain 0.2, 0.5, 1.0, 1.5 and 2.0% w/v of chloroform. Treat each solution in turn in the following way. Transfer the solution (0.1 ml) to a flask approximately 1 ml in volume and attach to an evacuated gas cell (Fig. 60). Open tap A and carefully heat the flask for 2 min at 150° in oil bath. With the flask still in the bath close tap A. Insert the gas cell into the sample beam and record the absorption spectrum over the range 820–700 $cm^{-1}$. Using the base-line technique calculate the absorbance of each solution and plot the values against concentration of chloroform to construct a calibration curve.

Dilute the sample (5 ml) to 25 ml with ethylene glycol and mix well. Treat 0.1 ml of solution as described above and determine the percentage w/v of chloroform in the sample by reference to the calibration curve. Convert the result to percentage v/v by dividing by the weight per ml of chloroform (1.48).

The apparatus (Fig. 60) is a normal gas cell of about 150 ml volume and reproducible results are obtained by using the same cell, flask and pipette for each determination. An internal standard is not necessary providing extreme care is taken in using the pipette.

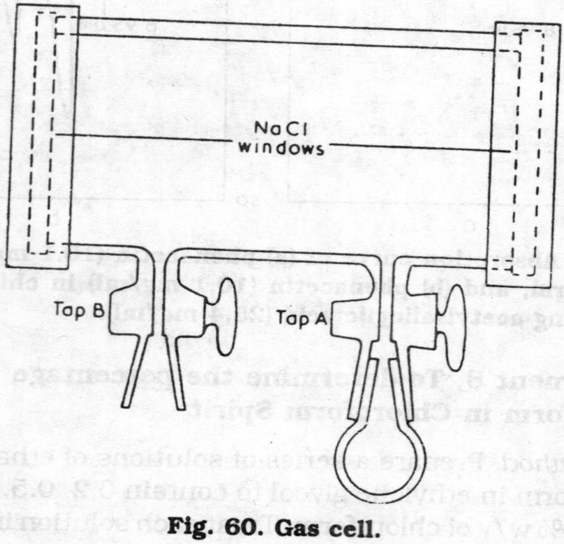


**Fig. 60. Gas cell.**

Tap B (Fig. 60) remains closed in this determination but when aqueous solutions are examined a small flask containing phosphorus pentoxide may be used to maintain a low level of water vapour during the determination.

### Cognate Determinations

*Compound Chloroform and Morphine Tincture*
*Chloroform Liniment*

### Experiment 9. To determine the percentage of methanol in Orciprenaline Sulphate

The determination of solvent residues in solid chemicals is a normal part of quality control and in included in such terms as loss on drying or volatile matter. Ordinary drying processes do not, however, guarantee complete elimination of water or organic solvent and a number of examples exist where specific methods of determination are applied—water by titration with Karl—Fischer reagent and acetone, dichloromethane, isopropanol and methanol by GLC.

Some of these are readily determined by vapour phase infrared absorption with sample weights considerably less than those required for GLC. The infrared spectrum also identifies the solvent.

*Method.* Accurately weight the sample (about 0.03 g) into the small bulb (Fig. 60), add diethylene glycol (0.2 ml) and attach to an evacuated gas cell. Complete the determination as described for chloroform (above) but using aliquot portions of methanol in diethylene glycol for calibration purposes.

*Note.* Methanol (1mg) gives a suitable absorbance.

### Cognate Determinations

*Colchicine* (0.1g) for determination of ethyl acetate or chloroform.

*Novobiocin Calcium* (0.15 g) for determination of acetone.

*Novobiocin Sodium* (0.15 g) for determination of acetone.

Commercial batches of Novobiocin slats contained acetone as solvent but this is not stated in the Pharmacopoeia which presumably allows other solvents.

*Streptomycin Sulphate* (0.15 g) for determination of methanol.

*Warfarin Sodium* (Clathrate) (0.1 g) for determination of isopropanol.

**Experiment 10. To determine the percentage of menthol in Peppermint Oil using the near infrared region of the spectrum**

*Method.* Prepare standard solutions of menthol in carbon tetrachloride to contain 0.5, 1.0, 1.5, 2.0 and 2.5% w/v of menthol. Measure the absorption curve of each solution in 4 cm glass or quartz cells over the region 8000–7000 $cm^{1}$. Calculate the absorbance at the peak absorption (about 7200 $cm^{-1}$) using the base-line technique and construct a calibration curve.

Accurately dilute the peppermint oil with carbon tetrachloride to give a concentration of menthol of about 1.5% and determine the absorbance at about 7200 $cm^{-1}$ as described above. By reference to the calibration curve, calculate the percentage w/v of menthol in the oil and finally convert to percentage w/w by using the weight per ml of the oil.

The near infrared region of the spectrum is becoming of increasing interest with the advent of sensitive detectors. The spectra consist mainly of overtones of the fundamental bands and are less intense than those of the latter. Nevertheless, because solvents can be used in cells of long path length, the region has potential value for determination of those groups which show reasonable absorption. Fig. 61 illustrates the groups which may be determined and the regions of absorption.

The assay for menthol depends upon the determination of the hydroxyl group, and is therefore less specific than an assay which involves absorption due perhaps to a skeletal vibration characteristic of menthol. The result of the assay should be similar to that obtained by the Pharmacopoeial method.

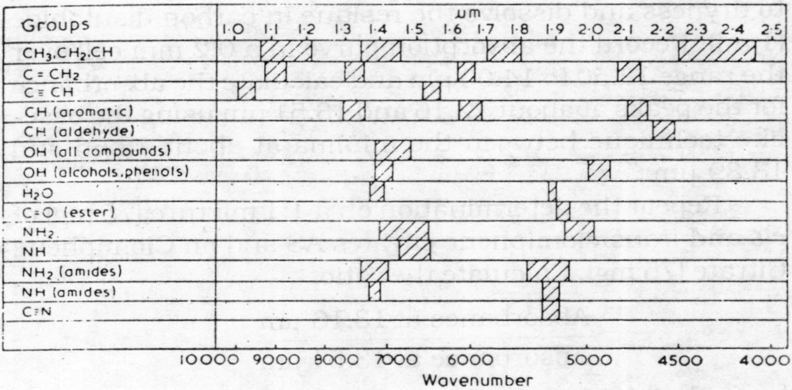


**Fig. 61. Characteristic absorptions in near infrared.**

## Experiment 11. To determine the cis-trans isomer ratio in Clomiphene Citrate

The shape of molecules often plays a significant role in the medicinal action of many compounds and if isomers are formed in the synthesis it may be important to include in the specification a test for the relative proportions of one to the other. This control ensures uniformity of composition of from batch to batch and one method is described for the *cis*-isomer of Tranylcypromine under Gas Chromatography. It is well known that *cis*- and *trans*-substituted double bonds have slightly different infrared absorption bands in the 13 µm region and this feature is the basis of the present determination.

*Method.* Dissolve *trans*-clomoiphene citrate AS (22.5 mg) and *cis*-clomiphene citrate AS (52.5 mg) in water (10 ml) in a separator and add sodium hydroxide solution (5%, 1 ml). Extract the mixture of bases with ether (3×10 ml) and wash the combined ether extracts with water (2×10 ml). Dry the ether with $Na_2SO_4$ (anhydrous), filter, evaporate to dryness and dissolve the residue in carbon disulphide (1 ml). Record the absorption curve in a 0.2 mm cell over the range 12.50 to 14.00 μm and calculate the absorbance for the peaks at about 12.16 and 13.51 μm using the base-line technique between the minima at about 12.66 and 13.89 μm.

Repeat the determination on a 1:1 mixture (75 mg) of *cis* and *trans*-clomiphene citrates AS and on Clomiphene Citrate (75 mg). Calculate the ratio

$$\frac{\text{Absorbance at } 13.16\ \mu\text{m}}{\text{Absorbance at } 13.51\ \mu\text{m}}$$

for each determination and confirm that the ratio for the sample lies between the ratios for the standards indicating that the sample contains 50–70% *cis*-clomiphene citrate.

**Experiment 12. To determine the percentage of thiotepa $C_6H_{12}N_3PS$ in Thiotepa Injection**

The preparation consists of sterile powder of Thiotepa, Sodium Bicarbonate and Sodium Chloride in a sealed container to which Water for Injection is added before use.

*Method.* Carry out the determination of Uniformity of Weight and from the mixed contents weight a quantity containing the equivalent of 75 mg of Thiotepa into a sinter-glass crucible. Extract with carbon disulphide (3 × 5 ml) (Note) and allow to filter through into a small flask. Evaporate the filtrates to low volume, transfer quantitatively to a 10 ml volumetric flask and dilute to volume with carbon disulphide. Measure the absorbance in a 0.1 mm cell over the region 9–12 μm and calculate the value at the maximum at about 10.75 μm using the base-line technique. Use Thiotepa AS treated in the same way for comparison and calculate the content of $C_6H_{12}N_3PS$ in the average contents of one ampoule.

*Note.* The utmost care should be exercised in the handling of carbon disulphide because of its toxicity and inflammability.

## Experiment 13. Determination of Cyclophosphamide in Cyclophosphamide Tablets

The purpose of this experiment is to illustrate the use of the potassium chloride disc in quantitative work by means of an internal standard. It has a great degree of specificity and, although rather long, is still more rapid than the pharmacopoeial method.

The internal standard must necessarily have few absorption bands and ferric thiocyanate meets this requirement. The method is described in the National Formulary XII (1965) but is incorrect in the USP XVIII in that the equivalent of 4 ml of internal standard solution is used for the tablets and reference is made to a standard curve prepared from 5 ml of internal standard solution.

*Internal Standard Solution.* Extract a mixture of ferric ammonium sulphate (2.63 g) and ammonium thiocyanate (26.3 g) in water (250 ml) with ethyl acetate (6 × 100 ml). Dry the extracts ($MgSO_4$, dried) and filter. Make up to 1000 ml with ethyl acetate.

*Standard Curve.* Use three quantities of Cyclophosphamide (50, 100 and 150 mg) accurately weighted and treat each as follows.

Dissolve in water (50 ml), extract with chloroform (3 × 10 ml) collecting the extracts in a 50 ml volumetric flask. Add the internal standard solution (4 ml) and dilute the volume with chloroform. Transfer the solution (1.00 ml) to a mortar containing potassium chloride (500 mg) and evaporate the solvent with the aid of a current of air and finally a vacuum (2 min). Prepare a disc from the dried powder and record the infrared absorption spectrum.

Calculate the absorbance (Note) of the internal standard band at 4.9 μm and the Cyclophosphamide band at 9.5 μm by the base-line technique. Calculate the ratio $A_{9.5}/A_{4.9}$ and construct a calibration curve.

*Method.* Treat the finely powdered tablets, accurately weighted (equivalent to about 350 mg Cyclophosphamide) in a sinter-glass crucible with chloroform as described for Phenobarbitone. Tablets, Transfer the filtrate to a 200 ml volumetric flask, add the internal standard solution (16.0 ml) and dilute to volume with chloroform. Proceed as described for the standard curve from the words 'Transfer 1.00 ml.. 'Calculate the ratio $A_{9.5}/A_{4.9}$ and read off the concentration of Cyclophosphamide from the curve. Calculate the content of Cyclophosphamide in the tablets.

*Note.* The term 'absorbance' instead of 'extinction' is used throughout this Chapter in conformity with current practice.

## Routine Maintenance

### Demountable Cells

The cells for use in infrared spectrophotometry require considerable care in storage and handling, as the sodium chloride or potassium bromide windows are fogged by traces of moisture and are easily scratched. A small cabinet fitted with a simple heater, e.g., a 60 watt lamp and a dish of silica gel as drying agent should be used for storing all cells, windows or plates, and relevant apparatus such as the die for preparing potassium chloride discs. The cabinet and contents are thereby maintained at about 10–20° above room temperature and there is little risk of fogging from atmospheric moisture.

In spite of all precautions, however, the windows eventually become opaque to visible radiation, and it is necessary to repolish them if the transmission of infrared radiation is much reduced by scatter. A polishing kit can be purchased which normally consists of a glass plate with a coarse and finely ground surface, a polishing cloth stretched on a firm base, polishing powder, e.g., jeweller's rouge, and an optical flat. They repay their cost in a short time as all types of defect, e.g., fogged or scratched surfaces in the windows, may be remedied.

The following system, however, has proved satisfactory in obtaining serviceable widows from sodium chloride 'blanks' within 6 to 7 min. Blanks offer a considerable saving in cost over the price of polished windows and they normally occur with surfaces similar to those obtained from the glass roughing and smoothing laps of a polishing kit.

*Materials.* Sodium chloride blanks
Two polishing cloths
Two thick glass or wooden bases on which the polishing cloths are firmly spread
Polishing powder
Ethanol containing a few crystals of sodium chloride.

*Method.* To one of the polishing cloths add a small amount of powder and saturate with ethanol. Place the blank over the powder and move it with firm pressure, rabidly, in a linear motion (Note). At short intervals rotate the blank and continue the polishing using more powder and ethanol. Examine the blank after one or two minutes by lightly drawing it over a clean portion of the cloth and continue the treatment if the surface is not smooth and transparent. Complete the polishing of the one side of the blank with rapid movements on the second clean and dry polishing cloth. Repeat for the other side of the disc.

*Note.* Some polishers prefer circular motion but this is a matter of personal choice.

This method will not remove deep scratches for which ground glass surfaces or very fine sand paper are necessary. Gloves are not essential for handling sodium chloride blanks.

## Sealed Cells

These consist of two sodium chloride or potassium bromide windows separated by a lead spacer, the whole being held together in a metal frame along with protective washers. Unless the transmission of the cell has deteriorated appreciably and the cell is in obvious need of renovation it is unwise to attempt to dismantle it. However, teflon spacers instead of lead spacers are now being used, and this simplifies considerably the dismantling and cleaning of cells.

*Method.* Note the construction of the cell, isolate the window assembly and carefully separate the two windows and lead spacer. These are normally firmly stuck but the insertion of a razor blade into the join and *gentle* leverage at each corner will generally separate them. Peel off the lead spacer and polish the two windows in the normal way.

# 2

# Raman Spectroscopy

## Raman Scattering

All molecules scatter light. Scattering occurs when a photon interacts with a molecule, but no absorption occurs. The electric field of the light pertubes the electron distribution of the molecule momentarily, but no transition occurs. Because the molecule does not go from one stationary state (ground-state energy level) to another (excited-state) energy level), there are no selection rules. Figure 1 indicates scattering occurring by a transition to virtual state (nonstationary state); the molecule immediately returns to the ground electronic state and the photon is scattered. If there is no change in energy of the scattered photon, the scattering is elastic; it is called *Rayleigh scattering*. The intensity of the scattered light is proportional to the square of the *polarizability* of the molecule, and if the wavelength is not too close to an absorption band, the scattering is inversely proportional to the fourth power of the wavelength. This wavelength dependence explains blue skies and red sunsets, although for the best red sunsets the scattering particles are dust, not molecules, and the wavelength dependences is more complicated.

The polarizability, $\alpha$, is a measure of how easy it is to induce a dipole in a molecule by applying an electric field. The polarizability tells us how easy it is to distort the distribution of electrons of the molecule. The quantitative definition is

$$\mu_{ind} = \alpha E$$

where $\mu_{ind}$ is the dipole induced and $E$ is the electric field. The usual units for $\alpha$ are $cm^3$. The electric field is then in esu-volt $cm^{-1}$ and $\mu_{ind}$ is esu cm. In SI units $\alpha$ is in $m^3$ and

$$\mu_{ind} = 4\pi\varepsilon_0 \alpha E$$

with $\mu_{ind}$ in coulomb m and $E$ in volt $m^{-1}$.

In Raman scattering the molecule returns to a different energy level after interaction with the light. In Fig. 1 the final state is an excited vibrational level of the ground electronic state. This is the most common occurrence in Raman scattering. The scattered photon has a longer wavelength (lower energy) than does the incident photon. The scattered photon can have a shorter wavelength if the molecule was originally in an exicted vibrational state and returned to the ground vibrational state. In either case the scattering is inelastic; there has been energy transfer between the molecule and the light. The selection rule for Raman scattering is that the polarizability of the molecule must change with the vibration in order to have a transition to a different energy level. Thus $N_2$ and $O_2$ will show vibrational Raman spectra, because their vibrations cause a change in polarizability. For $CO_2$ its asymmetric stretch will cause no change in polarizability, but its symmetric stretch and its bend will. We see that some vibrations can be measured using Raman, but not infrared; some can be measured using infrared, but no Raman; and some can be measured both ways.

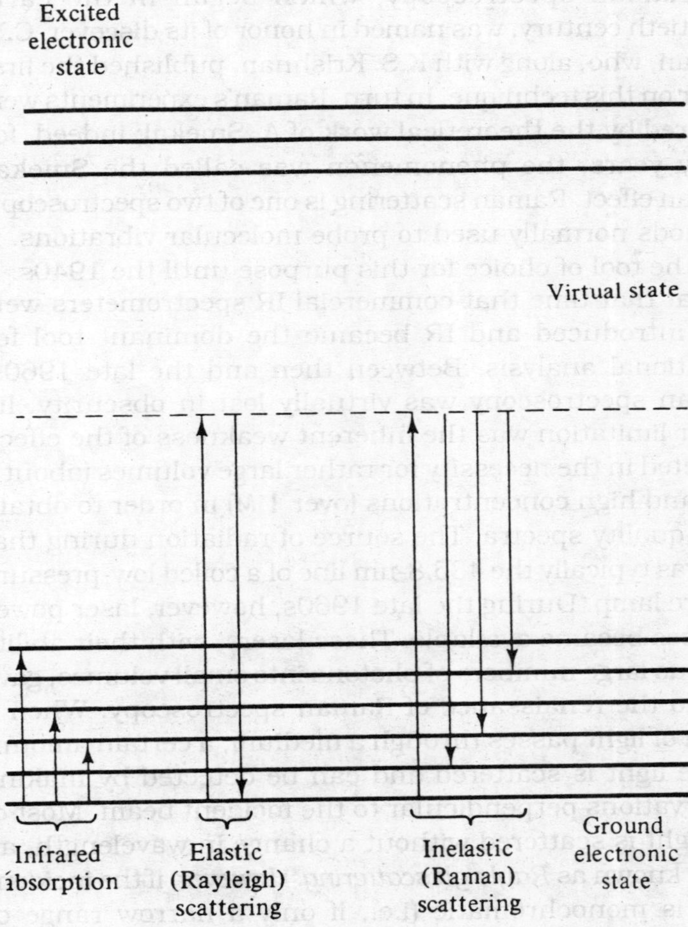


**Fig. 1. Energy-level diagram for infrared absorption, elastic (Rayleigh) scattering, and inelastic (Raman) scattering. Transitions between vibrational energy levels of the ground electronic state absorb infrared radiation; not all transitions are allows. For elastic scattering the frequency of the incident light is equal to the frequency of the scattered light. For inelastic scattering the frequency of the scattered light is different from that of the incident light. The virtual state shown can have any energy; the scattering of light occurs in any region of the spectrum.**

Raman spectroscopy, which began in the early twentieth century, was named in honor of its discover, C.V. Raman, who, along with K.S. Krishnan, published the first paper on this technique. In turn, Raman's experiments were inspired by the theoretical work of A. Smekal; indeed, for many years, the phenomenon was called the Smeka–Raman effect. Raman scattering is one of two spectroscopic methods normally used to probe molecular vibrations. It was the tool of choice for this purpose until the 1940s. It was at that time that commercial IR spectrometers were first introduced and IR became the dominant tool for vibrational analysis. Between then and the late 1960s, Raman spectroscopy was virtually lost in obscurity. Its major limitation was the inherent weakness of the effect, reflected in the necessity for rather large volumes (about 7 mL) and high concentrations (over 1 *M*) in order to obtain high-quality spectra. The source of radiation during that era was typically the 435.8-nm line of a coiled low-pressure Hg arc lamp. During the late 1960s, however, laser power sources became available. These lasers, with their ability to focus large numbers of photons into small volumes, gave rise to the renaissance of Raman spectroscopy. When a beam of light passes through a medium, a certain amount of the light is scattered and can be detected by making observations perpendicular to the incident beam. Most of the light is scattered without a change in wavelength, an effect known as *Rayleigh scattering.* However, if the incident light is monochromatic (i.e., if only a narrow range of wavelength is represented), a small amount of the scattered light may have either higher or lower wavelengths than the original light. The spectrum consists then of lines of both longer and shorter wavelengths than the incident wavelength. This effect was first observed in 1928 by the Indian physicist Sir Chandrasekhara Venkata Raman (1888–1970) and his co-worker Sir Kariamanikkam Srinivasa Krishnan (1898–1961) and is known as the **Raman effect.**

The Raman effect is a result of the interaction that occurs between a molecule and a photon. If a photon had a perfectly elastic collision with the molecule, it would be scattered with no change of frequency. The radiation would then be emitted in all directions, with an intensity inversely proportional to the fourth power of the wavelength. Because of this dependency, shorter wavelengths are more prominent in scattered light and particles suspended in a gas are rendered visible by a light beam.

On the other hand, in an inelastic collision, energy is exchanged between the molecule and the incident photon, according to quantum rules. The selection rule for rotational transitions in Raman spectroscopy is

$$\Delta J = 0, \pm 2 \tag{1}$$

The reason for this is that Raman scattering is a two-photon process, with one photon going in and one coming out, and as a result the angular momentum of the photons can remain unchanged or change by two units. If enough energy is available, a vibrational transition may occur, and the selection rule is then the same as for the infrared spectra of anharmonic oscillators; namely Dv = ±1, ±2,... Generally only the lowest vibrational level is occupied to any extent and the transition $\upsilon = 1 \leftarrow \upsilon = 0$ with rotational states superimposed gives the strongest Raman band.

## Theory of Raman Spectroscopy

### Rayliegh Scattering

When a molecule is placed in an electric field, the positive and negative charges are pulled to opposite sides. Thus a dipole moment is induced in the molecule. This happens even when the molecule has a permanent dipole moment.

The induced dipole moment $\mu_i$ is given by

$$\mu_i = \alpha^* \tag{2}$$

at moderate field strengths. The proportionality constant

$\alpha$ is called the **polarizability.** Consider now a molecule subjected to radiation of frequency $\omega_0$. Since the electric field oscillates according to

$$\mathscr{E} = \mathscr{E}^\circ \cos \omega_0 t \qquad (3)$$

the induced dipole moment of the molecule also oscillates. The induced moment therefore is given by

$$\mu_i = \alpha \mathscr{E}^\circ \cos \omega_0 t \qquad (4)$$

According to classical electromagnetic theory, an oscillating dipole radiates (scatters) at the frequency of oscillation. The scattering at the frequency of impinging radiation is called **Rayleigh** or **elastic scattering.**

## Raman Effect

We assumed in the above discussion that polarizability does not vary with time, but this is not true for molecules. We know that molecules rotate and vibrate with characteristic frequencies. Since induced moments depend on bond lengths and molecular orientation, the polarizability of a molecule changes as it rotates and vibrates.

Let us consider one particular frequency to illustrate the effect of changing polarizability. If $\omega_M$ is the frequency of rotation or vibration, the polarizability of the molecule varies with time according to the equation

$$\alpha = \alpha_0 + \alpha' \cos \omega_M t \qquad (5)$$

Therefore, the induced moments also vary with time:

$$\mu_i(t) = \alpha_0 \mathscr{E}^\circ \cos \omega_0 t + \alpha' \mathscr{E}^\circ \cos \omega_0 t \cos \omega_M t \qquad (6)$$

With the aid of a standard trigonometric relation, we can express this result in an alternative form :

$$\mu_i(t) = a_0 \mathscr{E}^\circ \cos \omega_0 t$$

$$+\tfrac{1}{2}\alpha' \ \mathscr{E}^\circ[\cos(\omega_0 - \omega_M)t + \cos(\omega_0 + \omega_M)t] \qquad (7)$$

We see from Eq. (7) that the induced dipole oscillates at the frequencies $\omega_0$ (Rayleigh scattering), $(\omega_0 - \omega_M)$, and $(\omega_0 + \omega_M)$. As a consequence, the molecules scatter radiation at these frequencies. The **inelastic scattering** at the sum and difference frequencies is known as the **Raman effect.** Fig. 2 illustrates the mechanism of Raman scattering.

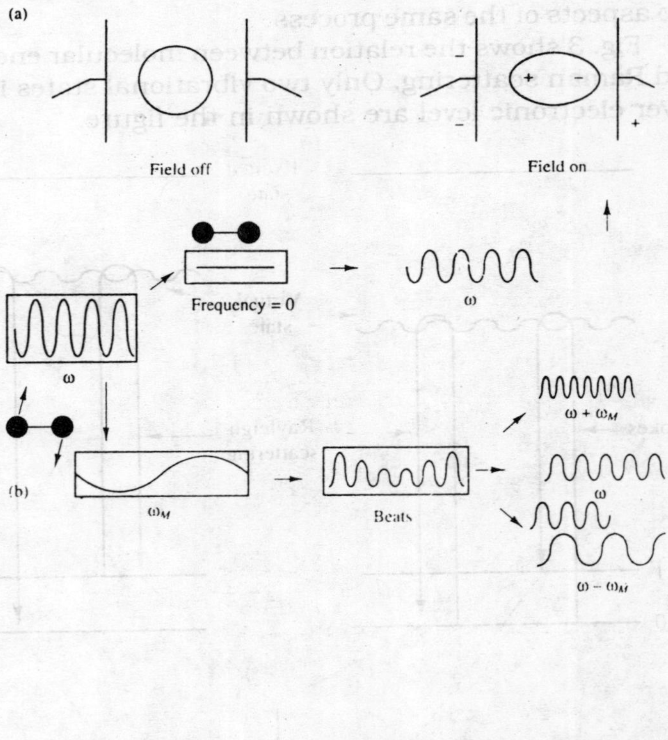


**Fig. 2. (a) A spherically symmetric charge distribution with no dipole moment is distorted to become a dipolar charge distribution when a field is applied. (b) When a molecule has only static polarizability (the frequency of motion is zero), it scatters the frequency of incident radiation as indicated at the top (Rayleigh scattering). If the molecule undergoes any motion (rotational, vibrational, or electronic at frequency $\omega_M$, it scatters radiation at ω (Rayleigh scattering) and ($\omega - \omega_M$) and ($\omega + \omega_M$) (Raman scattering).**

Let us examine the relation between Raman scattering and the beats. The mechanism for both these phenomena is the same, namely, the superposition of waves of different frequency. If we monitor the intensity of radiation as a function of time, we will observe the optical beats. If, on the other hand, we monitor the intensity of radiation as a function of frequency, we will observe scattering at different frequencies. Inelastic scattering and beat phenomena are two aspects of the same process.

Fig. 3 shows the relation between molecular energies and Raman scattering. Only two vibrational states in the lower electronic level are shown in the figure.

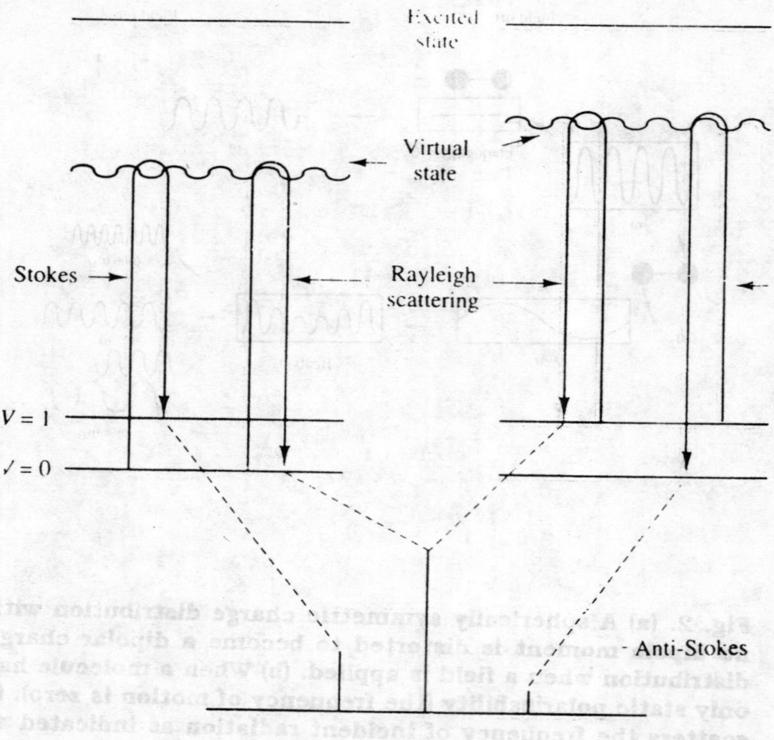


**Fig. 3. Relation among Rayleigh, Stokes and anti-Stokes Raman scattering.**

In the Raman studies the sample is exposed to radiation of a frequency not absorbed by the molecules. A Raman spectrum consists of weak scattered lines on both sides of an intense Rayleigh peak. Molecules in the $\upsilon = 0$ state scatter radiation at $(\upsilon_0 - V_M)$ ($\upsilon_M$ is the vibrational frequency). The resulting peak, which appears on the low-frequency side of the Raleigh peak, is called the *Stokes line.* The peak on the high-frequency sides of the Rayleigh peak is called the *anti-Stokes line.* This peak is due to scattering by the molecules in the $\upsilon = 1$ states. Since the numbers of molecules in the $\upsilon = 1$ state are lower than those in the $\upsilon = 0$ state, the anti-Stokes line is weaker than the Stokes line. Molecules gain energy in Stokes transitions and lose energy in anti-Stokes transitions.

The selection rules for Raman scattering are

| | |
|---|---|
| $\Delta v = 0$ | (pure rotational transitions) |
| $\Delta v = \pm 1$ | (ro-vibrational transitions) |
| $\Delta J = \pm 2$ | (linear molecules) |
| $\Delta J = \pm 2$ | ($K = 0$; symmetric top molecules) |
| $\Delta J = \pm 1, \pm 2$ | ($K = 1$; symmetric top molecules) |

It is not the dipole moment of the molecule but a change in the polarizability that determines whether a particular vibration is Raman-active or not. Figs. 4 and 5 give examples of Raman spectra.

We have considered above the classical theory of the Raman effect. However, serious study of this effect and related topics, such as coherent scattering process and multiphoton transitions, requires quantum theory. Even though the mathematical aspects of the theory are beyond the scope of this book, we can still appreciate the quantum explanation of the Raman effect.

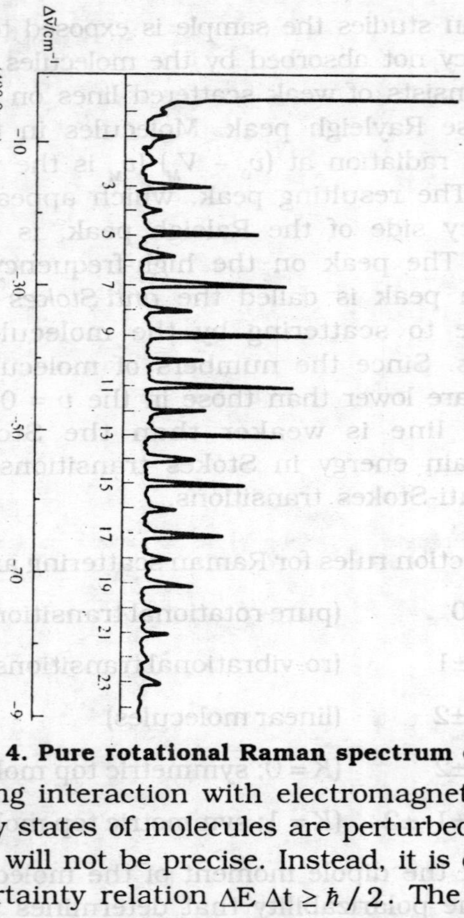


**Fig. 4. Pure rotational Raman spectrum of $^{19}F_2$.**

During interaction with electromagnetic radiation, the energy states of molecules are perturbed. The energy of a state will not be precise. Instead, it is controlled by the uncertainty relation $\Delta E \, \Delta t \geq \hbar / 2$. The shorter the time of interaction, the greater the energy spread for a state. These nonstationary states (i.e., those whose energy cannot be precisely defined because of uncertainty) are called **virtual states.** According to quantum theory, the effect of nonabsorbing radiation is to bring molecules into trasient virtual states from which they inelastically scatter radiation and return to a stationary state. We cannot describe the scattering process in the language of absorption and emission since scattering does not involve transitions between stationary states.

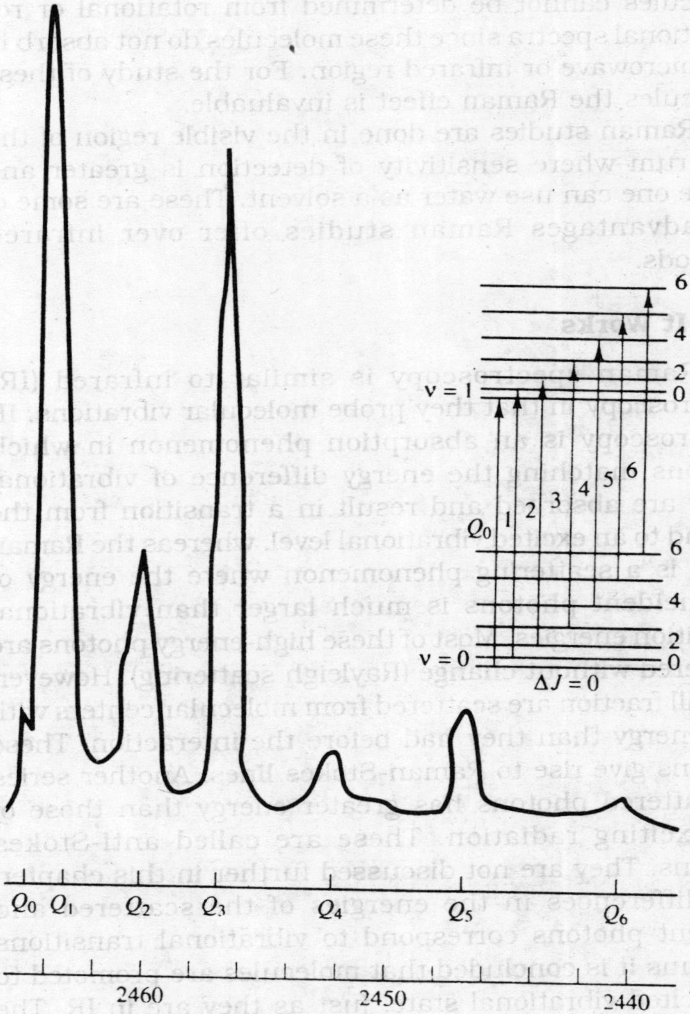


**Fig. 5. Vibrational Raman spectrum of $^3H$-$^3H$.**

Raman studies complement and enhance absorption studies in the infrared and far-infrared regions. For instance, the bond lengths of homonuclear diatomic molecules cannot be determined from rotational or ro-vibrational spectra since these molecules do not absorb in the microwave or infrared region. For the study of these molecules the Raman effect is invaluable.

Raman studies are done in the visible region of the spectrum where sensitivity of detection is greater and where one can use water as a solvent. These are some of the advantages Raman studies offer over infrared methods.

## How It Works

Raman spectroscopy is similar to infrared (IR) spectroscopy in that they probe molecular vibrations. IR spectroscopy is an absorption phenomenon in which photons matching the energy difference of vibrational levels are absorbed and result in a transition from the ground to an excited vibrational level, whereas the Raman effect is a scattering phenomenon where the energy of the incident photons is much larger than vibrational transition energies. Most of these high-energy photons are scattered without change (Rayleigh scattering). However, a small fraction are scattered from molecular centers with less energy than they had before the interaction. These photons give rise to Raman-Stokes lines. Another series of scattered photons has greater energy than those of the exciting radiation. These are called anti-Stokes photons. They are not discussed further in this chapter. The differences in the energies of the scattered and incident photons correspond to vibrational transitions and thus it is concluded that molecules are promoted to an excited vibrational state, just as they are in IR. The process is depicted schematically in Fig. 6, where it is labeled NR (normal Raman). The dotted line is called a virtual state and is little more than a mental crutch that allows us to visualize the process. It must be emphasized that this state does not exist. The information obtained in a Raman experiment can be quite different from that in an IR experiment because the quantum mechanical selection rules are different. In IR, an absorption can take place only if there is a change in dipole moment during the vibration; on the other hand, a change in polarizability (the ease with which the electron cloud is distorted by an external electric field) is required in order for Raman scattering to occur. This gives rise to some rather interesting and useful differences between IR and Raman spectroscopies. Homonuclear diatomic compounds such as $H_2$, $Cl_2$, or (in general) $X_2$ are IR-inactive but active in Raman. These are specific instance of the general rule that molecules with a center of symmetry exhibit a mutual exclusion effect. In this case, if a molecular vibration is active in IR, it is not in Raman and vice versa.

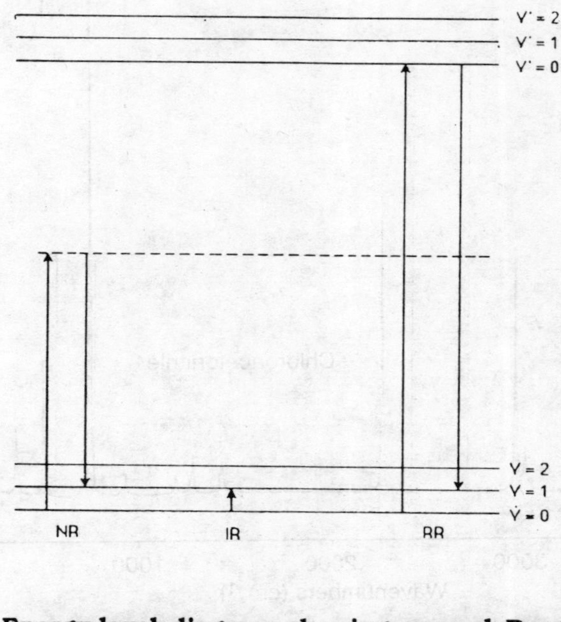


**Fig. 6. Energy-level diagram showing normal Raman (NR), resonance Raman (RR), and infrared (IR) energy transfer processes.**

Using IR and Raman together allows one to establish the existance of a center of symmetry quite easily. In fact, there is a loose rule that states that if a band is strong in Roman in IR it will be weak in Raman and if it is strong it will be weak in IR. Fig. 7 shows an example of this behavior.

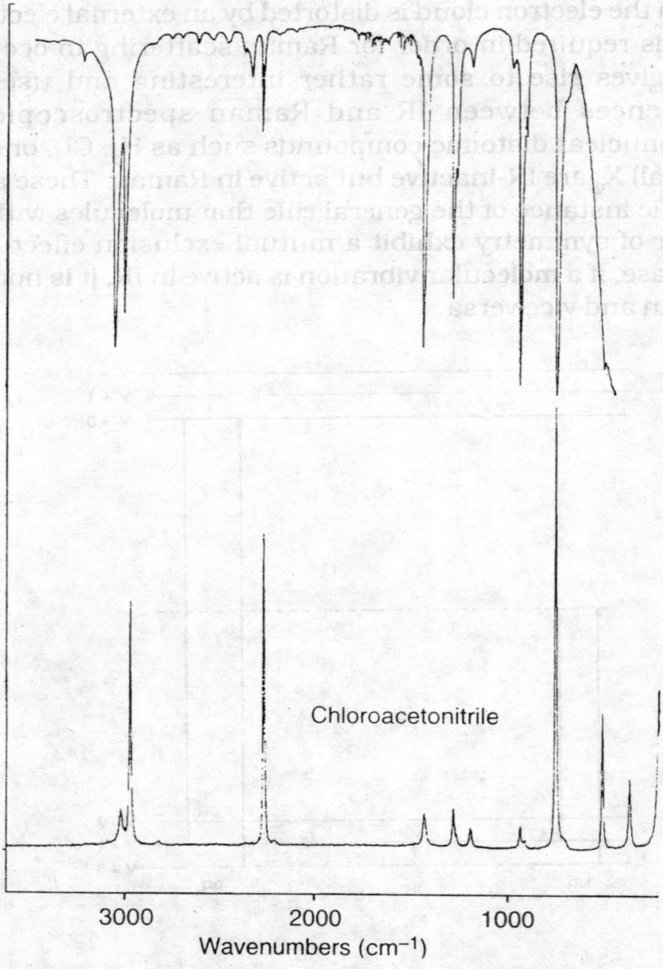


**Fig.7. Raman (lower trace) and IR (upper trace) spectra of alpha-chloroacetonitrile, Raman-excited with 488–m radiation.**

Note particularly the band due to the C–N stretching mode at about 2200 cm−1. All of this occurs because the IR activity of a vibration mode requires a change in dipole moment during the motion, but Raman activity requires a change in the polarizability. Thus, in a centrosymmetric molecule, the totally symmetric vibration does not give rise to a change in dipole and is IR-inactive. However, it can readily be shown that it does give rise to a polarizability change and consequently is Raman-active.

Raman scattering information is presented as an intensity-versus-wavelength shift in $cm^{-1}$ and can be conveniently compared to its IR counterparts. It is important to realize that the position of a given IR band in $cm^{-1}$ is identical to the wavelength shift of the Raman band because the same transition is being excited by both process. Thus, fundamental group frequencies from IR tables can be conveniently used in Raman studies to obtain structural information. The number of reference spectra available for Raman is limited as compared to IR. However, several tabulations are available.

As is true for IR, the number of fundamental vibrational modes in 3n–6 for nonlinear molecules and 3n–5 for linear molecules. The actual number of bands observed in the spectrum may be diminished in the same fashion as in IR. That is, some modes are not active and degeneracies may occur. However, in normal Raman overtones, combination and difference bands are rare. This results in spectra that are significantly simpler than their IR counterparts (Fig. 8).

A limitation of normal Raman spectroscopy is its inherent weakness. Solutions that are on the order of 0.5 *M* or greater are needed to obtain satisfactory spectra. However, if the energy of the incident photons falls within an excited state manifold (Fig. 6), then the signal due to excitation of particular vibrations may be enhanced as much as one-thousand-fold. This is the resonance Raman effect. It should be noted that the arrowheads corresponding to the energy of the incident photons fall between vibrational energy levels of the excited state. This was done to emphasize the difference between resonance Raman scattering and resonance fluorescence.

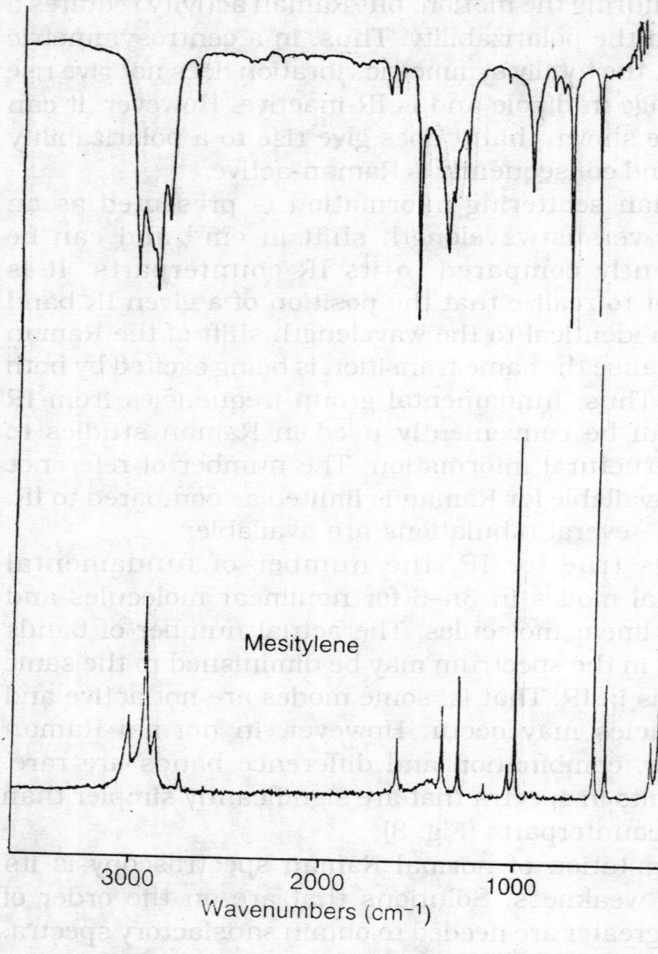


**Fig. 8. Raman (lower trace) and IR (upper trace) spectra of mesitylene.**

In the latter, an actual absorption and re-emission takes place that results in depolarized radiation. This is not so for Raman scattering, where the scattered radiation may be highly polarized. Because the signal intensity is $10^3$ times stronger than in normal Raman scattering, solutions on the m*M* scale may thus be used to produce satisfactory spectra. The spectra become even simpler than those obtained with normal Raman because only totally symmetric vibrations show this large signal enhancement. Fig. 9 dramatically shows this effect using various laser lines to excite the spectrum of $Fe(phen)_3^{2+}$. When the low-energy laser line is used, the spectrum of the compound is virtually unobservable. When energies within the region of the absorption maximum (510 nm) are used, strong clean spectra are obtained.

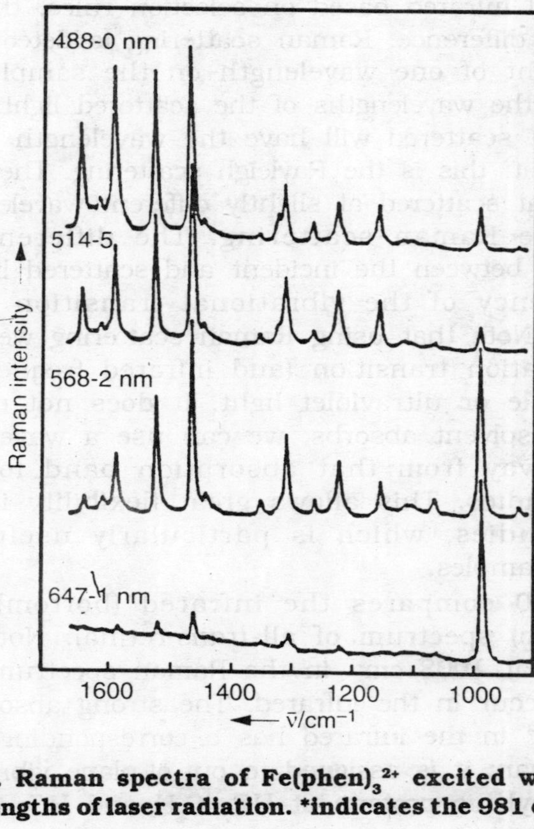


**Fig. 9. Raman spectra of $Fe(phen)_3^{2+}$ excited with various wavelengths of laser radiation. *indicates the 981 $cm^{-1}$ of $SO_4^{2-}$.**

Another limitation of normal Raman scattering is that it is much weaker than fluorescence emission. The latter may originate from the sample itself or some small impurity in the sample. Because the region of the electromagnetic spectrum where fluorescence occurs is always the same, but the Raman wavelength depends on the exciting line, it is possible to avoid the fluorescence problem by using an alternate exciting line. Recently, the advent of FT-Raman offered another solution to this problem. By exciting in the near IR, 1.06 μ with a Nd–YAG laser, we can avoid exciting the fluorescence in the first place.

In addition to the fundamental difference between Raman and infrared based on selection rules, there is a practical difference. Raman scattering is detected by shining light of one wavelength on the sample and measuring the wavelengths of the scattered light. Most of the light scattered will have the wavelength of the incident light; this is the Rayleigh scattering. There will also be light scattered at slightly different wavelengths; this is the Raman scattering. The difference in frequencies between the incident and scattered light is the frequency of the vibrational transition being measured. Note that using Raman scattering we learn about vibration transition (and infrared frequencies) using visible or ultraviolet light. It does not matter where the solvent absorbs; we can use a wavelength of light away from that absorption band for the Raman studies. This allows great flexibility in the Raman studies, which is particularly useful for biological samples.

Fig. 10 compares the infrared (bottom) and Raman (top) spectrum of all-*trans*-retinal. Note the large peak at 1008 $cm^{-1}$ in the Raman spectrum that does not occur in the infrared. The strong absorption at 966 $cm^{-1}$ in the infrared has a corresponding peak in the Raman; it is assigned to out-of-plane vibrations of *trans*-ethylenic protons at $HC_7{=}C_8H$ and $HC_{11}{=}C_{12}H$.

Deuterium substitution and synthesis of dimethyl analogs have led to a complete assignment of the vibrational spectra of retinal. Because the Raman spectra of visual receptors and bacteria which use retinal as a key chromophore are being studied in single cells, the changes of conformation of the retinal can be deduced during its biological function (Barry and Mathies, 1982).

Resonance raman spectra occur when the exciting light is in an electronic absorption band of the molecule. There can be gains in sensitivity for vibrations within the absorbing chromophore of a factor of 1000, or more. This allows one type of molecule to be studied preferentially in a very complicated mixture. Resonance raman has been applied to heme proteins and to rhodopsin.

Surface-enhanced Raman spectroscopy (SERS) is another recently developed technique for obtaining strong signals from small numbers of organic molecules. It was first reported in 1974 by Fleishman and colleagues. Samples adsorbed onto microscopically roughened metal surfaces or colloids give spectra as much as $10^6$ stronger than expected from normal Raman. Metals that are effective include gold, silver, and copper, although zinc and gallium have also shown the effect.

## How to Obtain Raman Spectra

Raman spectra are obtained by measuring the intensities and frequencies of scattered radiation originating from a sample that has been irradiated with a monochromatic source. The units on the abscisa are D $cm^{-1}$, where D indicates the difference in $cm^{-1}$ between the energy of the incident and scattered photons. These are the vibrational energies of the sample molecules. Typically the source is a laser; a common experimental setup for obtaining spectra is shown in Fig. 11.

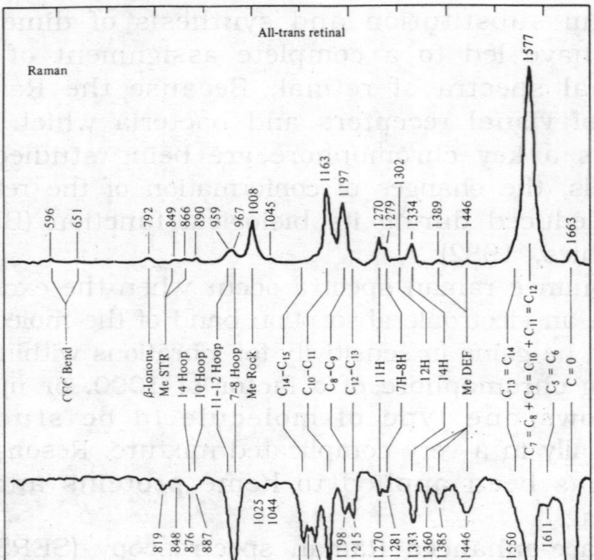


**Fig. 10. Raman and infrared spectra of all-trans-retinal in the region from 600 cm$^{-1}$ to 1600 cm$^{-1}$. The infrared absorbance is inverted to facilitate comparison with the raman. The raman spectrum was measured in $CCl_4$ using 676.4-nm excitation from a krypton ion laser. Solvent peaks were subtracted. The infrared spectrum was measured with a thin film deposited by evaporation from a pentane solution on a KBr window. All peaks have been assigned to particular vibrations of the molecule. Me STR, Me Roek, and Me DEF refer to the stretching, rocking and deformation of methyl groups; Hoop means out-of-plane motion of H atoms. Note that double bond vibrations occur at higher frequencies than single bond bivrations. (From B. Curry, I. Palings, A. Brack, J.A. Pardoen, P.P.J. Mulder, J. Lugtenburg, and R. Mathies, J. Phys. Chem. 1984, 88, 688–702. The figure was kindly supplied by Prof. Richard Mathies, University of California, Berkeley).**

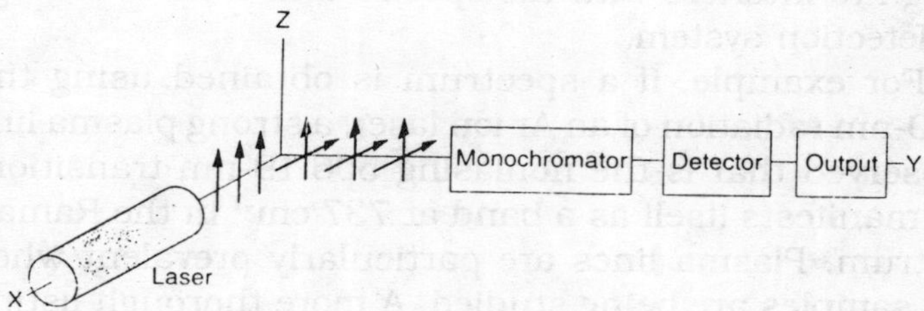


**Fig. 11. Standard 90° arrangement for the detection of Raman scattered radiation. The sample is located at the origin.**

## Dispersive Spectrophotometers

These have traditionally used a double-grating monochromator (sometime even a triple) in order to eliminate the intense Rayleigh scattering and the gratings normally have 1200 to 1800 grooves/mm. The Czeryny–Turner optical arrangement used by Spex Industries is whom in Fig. 12. These instruments have four slits that are manually controlled to select resolution and throughput. Recently developed filters, called supper notch filters, have eliminated the need for double monochromators in many applications. They can reduce Rayleight scattering to a point where spectra may be obtained to within 150 $cm^{-1}$ of the exciting line. However, studies requiring closer approach to the Rayleigh line still require the use of double monochromators. Some of the scattered radiation may be highly polarized and because the efficiency of the gratings depends on polarization, it should be randomized (scrambled) by passing it through a quartz wedge. Failure to do this can result in spectra in which the intensities of certain bands are anomalously low. It is desirable to pass the laser radiation used to excite the Raman scattering through a premonochromator in order to remove the plasma lines from the beam. These plasma lines originate from electronic transitions that, though not lasing, are strong enough to interfere with the spectra and even to damage the detection system.

For example, if a spectrum is obtained using the 488.0-nm radiation of an Ar ion laser, a strong plasma line is observed that is the nonlasing 506.19-nm transition. This manifests itself as a band at 737 $cm^{-1}$ in the Raman spectrum. Plasma lines are particularly prevalent when solid samples are being studied. A more thorough listing of these lines and where they appear in spectra may be found in the book *Laboratory Raman Spectroscopy* by Strommen and Nakamoto. It is important to note that fluorescent room lights give rise to at least one spurious line for each excitation wavelength. Thus, when using 488.0-nm excitation, a line is observed at 2180 $cm^{-1}$ and when using 514.5-nm excitation, the same line is observed at 1125 $cm^{-1}$.

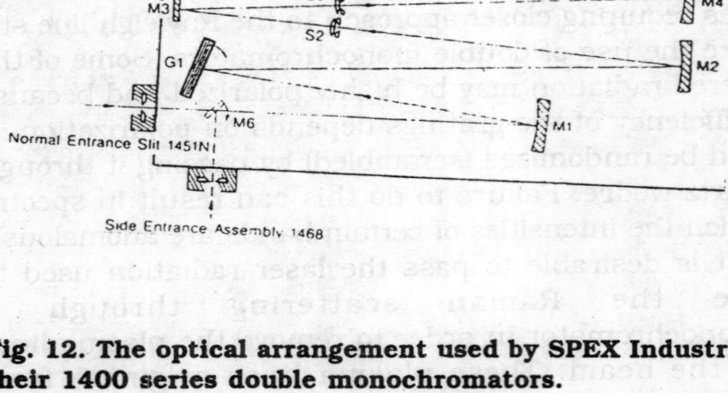


**Fig. 12. The optical arrangement used by SPEX Industries in their 1400 series double monochromators.**

Detection is most often accomplished by use of a thermometrically cooled photomultiplier tube (PMT). Dispersive spectra may also be obtained with photodiode array (PDA) detectors and charge transfer devices (CTDs). The PDA is the simplest and consists of an array of photodetectors. A photodetector can be classified as either a photoconductive or a photovoltaic detector. When operated photovoltaically, the photodetector responds to light and generates an electrical current proportional to the number of photons that strike its surface. When the photodetector is operated as a photoconductor, its electrical resistance is reduced proportionally to the number of photons that strike its surface. Reverse-biasing a photodiode (applying a positive voltage to the photocathode) is generally used on PDAs to increase response time at the expense of optical sensitivity. A PDA is normally used together with an image intensifier to increase the weak Raman signal.

CTDs can generally be divided into two types: charge-coupled devices (CCDs) and the charge-injection devices (CIDs). In both the CCD and the CID arrays, a photosite converts the incoming optical signal to charge, which is stored in the device (integrated). The CCD is a metal insulator semiconductor that can store charge in localized potential wells. A CCD array can be visualized by picturing the potential wells as a row of buckets (pixels) filled with charge. To read the information from a CCD array, the charge is transferred from the end pixel (bucket) to the readout electronics. The pixel that is adjacent to the end pixel then transfers its charge to the end pixel, which, in turn, transfers it to the readout electronics. This transfer to charge continuous sequentially from pixel to pixel until all of the pixels have transferred their charge to the readout electronics. The difference in types of CCD arrays are due mainly to matrix size and the method of reading out the pixels (column readout, row readout, or a combination of the two). There are two main disadvantages to CCD arrays. One is the limitation to sequential readout of the pixels, which causes these arrays to be slower than the PDA. The other is the possibility of blooming. Blooming is an undesirable effect that can be thought of as charge overfilling a pixel (bucket) and spilling into the adjacent pixel. Both of these limitations are addressed in the CID array. The CID array is a hybrid between discrete detectors and monolithic arrays that includes some signal processing on the photodetector substrate. CID arrays can be manufactured from materials that cannot be used in CCDs. It has the advantage of random access to pixels, and is relatively insensitive to blooming.

When selecting the type of array to use (PDA, CCD, or CID), one should consider the complexity of optics necessary to collect and focus the light signal onto the array, the speed of light detection necessary, and the sensitivity required. As a rule, a PDA requires more sophisticated optics and filtering to ensure that the appropriate portion of the light signal strikes only the desired photodetector with no optical crosstalk between elements. With a CCD or CID, this effect is less pronounced because a larger area is available on which to direct the optical signal. The speed of light detection necessary (the system bandwidth) depends on how fast the rise time of the optical signal is. A CCD is the slowest of the three arrays and the PDA has the potential to be the fastest. Sensitivity of the array is an issue if very weak signals are to be detected. Generally a PDA is more sensitive than a CCD or a CID because the CTD has an associated charge-transfer efficiency associated with it. These efficiencies can be high, but are still less than 1, making these devices less sensitive than a PDA. For overall ease of use, the CCD is probably best because it can be purchased relatively inexpensively, but it is the least flexible of the three arrays.

The choice of photosensitive material for both the PDA and CTD is dictated by the light wavelengths of interest.

Materials are available from the visible region to the IR region. However, it should be noted that PDAs and STDs require cryogenic cooling and are difficult to manufacture with any uniformity between photodetector elements.

## Fourier Transform Spectrometers

These instruments use Michelson interferometers for wavelength analysis and ND-YAG laser as sources. The principles of Fourier transform (FT) spectroscopy are discussed. The optical arrangement for the Perkin–Elmer System 2000R Near-Infrared FT-Raman Spectrometer is shown in Fig. 13. The radiation source emits at 1.06 μ and the scattered radiation may have wavelengths as long as 1.7 μ, a region that is unsuitable for PMTs, PDAs, and CTDs. In this region photoconductive devices function nicely as detectors. InGaAs and Ge are commonly used materials. They are often used at cryogenic temperatures to lower the noise level and thus raise the signal-to-noise ratio. However, this also reduces the range over which the detector is effective. For example, the cutoff for InGaAs at room temperature is approximately 3500 $cm^{-1}$, but at liquid nitrogen temperature it drops to 2900 $cm^{-1}$. The 1.06-μ radiation that is often used in FT-Raman systems is supplied by a continuous-wave (CW)Nd–YAG laser with output power ranging up to 3 W.

It is worth noting again that fluorescent room lights must be rigorously excluded from the optical bench because they rise to numerous spurious bands between 1700 $cm^{-1}$ and 3000 $cm^{-1}$.

## What It Does

Sampling techniques used in obtaining Raman spectra range from the extremely simple to the rather complicated. In contrast to IR, where water is an impediment due to its strong absorptions across the vibrational energy region, Raman scattering due to water is quite weak. Thus aqueous solutions may be easily examined and glass containers may be used.

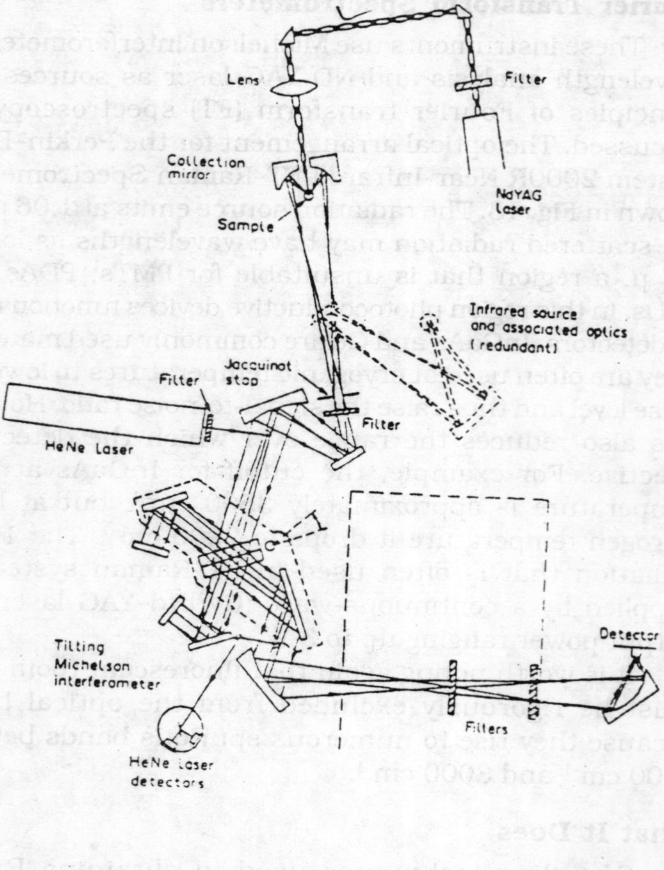


**Fig. 13. Optical design of the Perkin-Elmer 2000R Near Infrared FT-Raman Spectrometer.**

Low-frequency spectra, down to 50 $cm^{-1}$, are readily obtained and there is no need to use specialized instruments to change beamsplitters and sources or to maintain the sample in a vacuum, as is the case when probing low-frequency vibrational modes with IR instrumentation. The types of sampling devices used for resonance Raman differ from those used for normal Raman. The following are descriptions of a few of these sampling procedure as well as discussions of SERS and Raman microprobe experiments. More detailed descriptions and additional devices may be found in the monograph by Strommen and Nakamoto.

## Normal Raman

Liquid samples may be contained in glass capillary tubes, NMR tubes, cuvettes, and even beakers. In fact, if the bottle containing the sample is clear glass, the sample need not be removed in order to obtain a satisfactory Raman spectrum. When using glass capillary tubes only a few microliters are needed. This is because the scattering volume is limited by the diameter of the laser beam at the focal point. This diameter typically is less than 0.01 mm.

Solid samples are equally simple to deal with. Powders and microcrystalline or amorphous materials may be packed into melting-point capillaries or NMR tubes. Translucent solids may be mounted as-is at the focal point of the laser beam. For example, the spectrum of the handle of a screwdriver can be obtained by placing the handle in the beam. The location of the beam within the handle is not critical to the experiment.

Specialized sampling devices are well-documented in the literature. Temperature studies of Raman band positions and intensities can be done with a Harney–Miller cell. This cell holds the sample in a glass capillary that is bathed in a stream of nitrogen whose temperature is controlled and monitored. The lower limit of this cell is determined by the boiling point of nitrogen. Temperatures below this value may be reached with the aid of liquid He transfer cells or, more conveniently, with a closed-cycle. He refrigerator, which cools a Cu block on which the sample is deposited. Samples may be vaporized and condensed on the cold surface or they may be trapped (matrix isolation) in an inert gas matrix that is condensed with the sample. Spectra obtained in this fashion are free from intermolecular interactions and as a result are very sharp, often revealing details unavailable when the sample is restricted by other environments. High pressure diamond anvil cells similar to those used in IR experiments are also used in Raman spectroscopy.

## Resonance Raman

Liquid Raman must be handled differently because one must use radiation whose energy matches that of an absorption of the molecule. Given the high photon flux of the laser and the tightness of focus of the beam, local heating becomes a potential problem. In order to prevent both thermal lensing (which distorts the focus) and thermal decomposition, the laser beam must be prevented from irradiating the same volume for any significant length of time. Two general methods have been devised that minimize the local heating effect. In one of these the sample is placed in a round glass container with an optically flat bottom. The container is then rotate rapidly, forcing the sample to the sides. The difficulty with this approach is that the laser beam is brought up from the bottom and it is difficult to keep the laser and the container wall perfectly vertical and at the same time at the exact position of focus for the collection optics. A technique that avoids this problem uses a spinner that holds an NMR tube containing the sample. The beam is brought in on an angle and the scattering is observed at 135°.

Spinning devices such as these may be conveniently constructed from the spinner mechanisms of old NMR systems such as the Varian A60.

An alternative approach to the heating problem involves flowing the sample through a closed loop using a peristaltic pump. The loop contains a section of glass capillary tube that is positioned in the laser beam. The advantage of this method is that alignment is rather simple. One can align a capillary containing some strong scatterer such as toluene and then merely replace it with the capillary in the closed loop. The final focal point will be very close. The disadvantage of this method is that it generally requires a much larger amount of the sample than does the rotating cell technique.

Solid samples must also be rotated in order to reduce the possibility of localized heating. This can be done by using the NMR spinner technique described previously. An alternative method requires the preparation of a pellet that is rotated while the beam is focused on or near the edge of the pellet. The pellets are prepared in a fashion similar to KBr pelletes for IR spectra. Approximately 200 mg of KBr is ground and placed in a pellet press. Then 5000 to 7000 $lb/in.^2$ pressure is applied for approximately 3 min. At this point the press is opened and 100 mg of the ground sample is spread on top as evenly as possible. The sample may also be mixed with an internal standard such as $Na_2SO_4$ in order that comparisons may be made of relative intensity changes as a function of excitation wavelength. The press is reassembled and the pressure is again raised to 7000 $lb/in.^2$ for an additional 3 min. The pellet is then removed and mounted on the rotator at a 45° angle with respect to the laser beam. It should be noted that none of these techniques prevent photodecomposition, which occurs on a time scale that is much faster than rotation times.

## FT-Raman

This technique, which is virtually free of interference from fluorescence, can be used to examine a wide variety of samples. Dyes are conveniently examined and may be determined at extremely low levels as described in the following section on SERS. Its main drawback is the fact that water absorbs in the 1.0 µ region and the commonly used source, a CW ND–YAG laser, radiates energy at 1.06 µ. Thus, the advantage of being able to study aqueous solutions is eliminated. Because of the freedom from fluorescence interference and ease of sampling, FT-Raman spectroscopy has application as an industrial tool. It has been shown to be useful in the study of the morphology of polyethylene and is commonly used to obtain data on polyolefins (10). However, it must be noted that the great advances made in CCD detector technology have somewhat diminished the potential of FT-Raman. Currently available CCD detectors are useful out as far as 1 µ.

## Sers

For the SERS effect to be observed, the molecules must be approximately 50 Å or less from the surface. Two of the techniques by which this is accomplished are adsorption onto metal films and adsorption onto colloids. The latter has been put to rather ingenious use in combination with FT-Raman. Samples run on TLC plates are subsequently sprayed with a Ag colloid solution. The resulting admixture evokes the SERS effect. Using this technique, extremely small amounts of material may be studied. In Fig. 14, a Raman spectrum of a sample of rhodamine B was obtained in situ after a 40-ng spot on a TLC plate was sprayed with a silver colloid. A recent development in this field involves the use of laser ablation of metals to prepare SERS-active colloids. The pure metal is suspended in a solvent and is irradiated for 10 to 15 min with 1064-nm radiation from a Nd-YAG laser. The resulting colloidal suspension is mixed with the substrate to be studied. In this fashion, a Raman spectrum of $5.0 \times 10^{-6}$ *M* cytochrome C was obtained with 3 mW of 514.5-nm excitation.

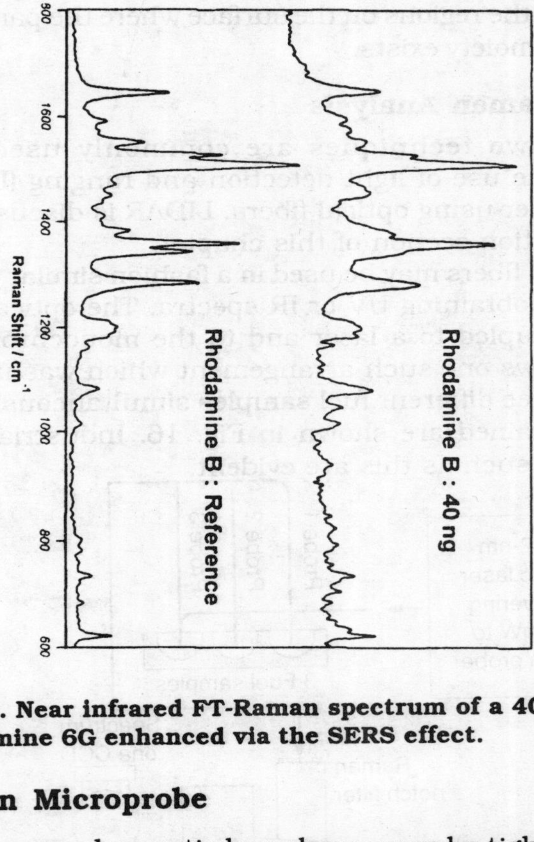


**Fig. 14. Near infrared FT-Raman spectrum of a 40-ng spot of Rhodamine 6G enhanced via the SERS effect.**

## Raman Microprobe

A monochromatic laser beam may be tightly focused through the use of a microscope objective, allowing the operator to examine regions on or within a sample, where the region of interest may be as small as 2.0 μ. Thus, areas of a corroded surface may be microscopically analyzed. Inclusions in minerals and meteorites may also be studied. The latter application is of particular interest to geochemists, who can determine the gas pressure within inclusions by monitoring the position of the 2917–cm$^{-1}$ Raman band, which is pressure-sensitive. With this information, the processes involved in the formation of the mineral may be ascertained. When examining surfaces, the instrument may be tuned to a given vibrational mode. This results in an image of the regions on the surface where the particular vibrational moiety exists.

## Remote Raman Analysis

Here two techniques are commonly used, one involving the use of light detection and ranging (LIDAR) and the other using optical fibers. LIDAR is discussed in the Application section of this chapter.

Optical fibers may be used in a fashion similar to that used when obtaining UV or IR spectra. The optical fiber probe is coupled to a laser and to the monochromator. Fig. 15 shows one such arrangement which was used to monitor three different fuel samples simultaneously. The spectra obtained are shown in Fig. 16. Industrial uses for a setup such as this are evident.

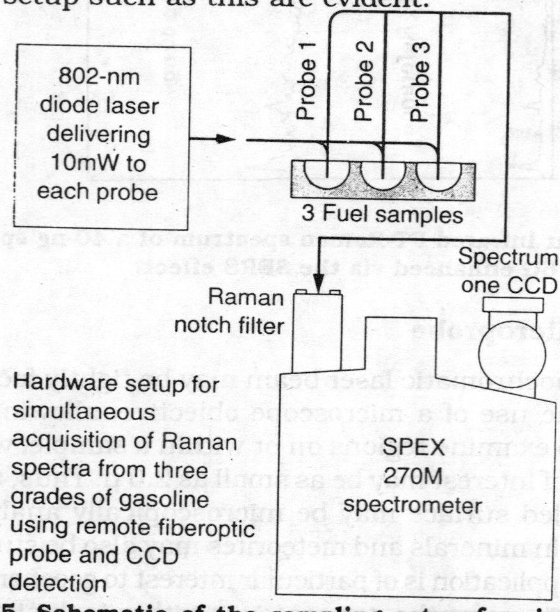


**Fig. 15. Schematic of the sampling setup for simultaneous determination of fuel samples.**

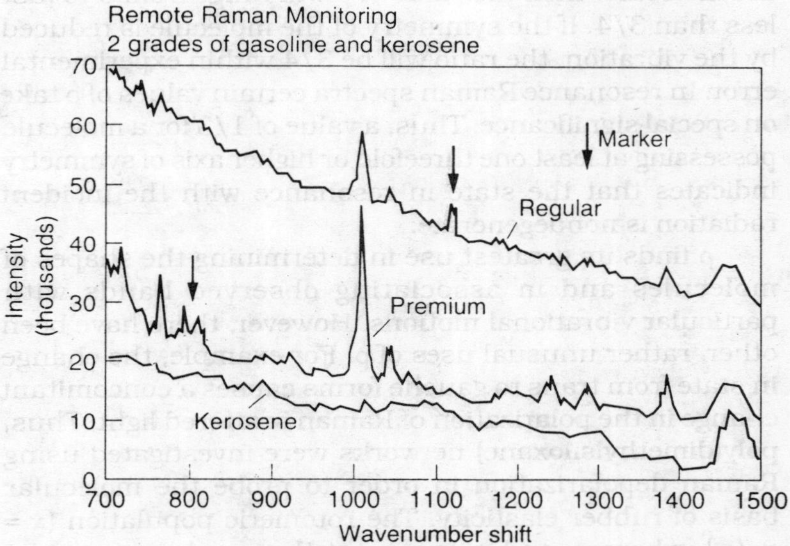


**Fig. 16. Raman spectra obtained using the experimental setup shown in Fig. 15.**

## Raman Depolarization Ratios

No discussion of Raman spectroscopy would be complete without a discussion of depolarization ratios. The depolarization ratio may be defined as

$$\rho = I_{\perp} / I_{//} \tag{1}$$

where $I_{\perp}$ = the intensity of scattered radiation whose plane of polarization is perpendicular to that of the incident radiation and $I_{//}$ = the intensity of scattered radiation whose plane of polarization is parallel to that of the incident radiation. Determination of $\rho$ is a rather simple matter. One merely records the spectrum with a piece of polarizing material (the analyzer) placed between the sample and the monochromator. The analyzer is positioned to pass one type of radiation and then rotated to pass the other.

If the vibrational motion preserves the symmetry of the molecule, then the value of ρ will range from 0 to just less than 3/4. If the symmetry of the molecule is reduced by the vibration, the ratio will be 3/4 within experimental error. In resonance Raman spectra certain values of ρ take on special significance. Thus, a value of 1/3 for a molecule possessing at least one threefold or higher axis of symmetry indicates that the state in resonance with the incident radiation is nondegenerate.

ρ finds its greatest use in determining the shapes of molecules and in associating observed bands with particular vibrational motions. However, there have been other, rather unusual uses of ρ. For example, the change in state from trans to gauche forms causes a concomitant change in the polarization of Raman scattered light. Thus, poly(dimethylsiloxane) networks were investigated using Raman depolarization in order to probe the molecular basis of rubber elasticity. The rotomeric population ($x = n_g/n_t$), where $n_g$ and $n_t$ represent the number of gauche and trans molecules, respectively, was obtained from the depolarization ratios of the 2907 $cm^{-1}$, $CH_3$ stretching vibration, and the 491 $cm^{-1}$ Si–O backbone stretching mode using the relationship

$$\rho = \rho_g \frac{x}{1+x} \tag{2}$$

Here again, the subscript *g* refers to the gauche form of the molecule. A more complete discussion of the origins and used of ρ is available.

## Analytical Information

It should be noted that Raman spectroscopy is a single-beam technique where internal standards are normally but not always used if quantitative results are required. The intensity of a Raman signal may be represented by Eq. (3).

$$I = K\nu^4 \oint C \tag{3}$$

Here $I$ represents the intensity of the Raman band, $K$ is a constant the includes the laser power at the sample, $\upsilon$ is the frequency of the scattered radiation, $\mathcal{J}$ is a scattering coefficient that is characteristic of each Raman band, and $C$ is the concentration of the analyte under consideration. A working curve is constructed by obtaining the spectra of a series of concentrations of the desired analyte, each containing the same amount of an internal standard. The internal standard should be chosen so that it has a mode relatively close in frequency to a mode of the analyte under consideration. If this is done, a series of relative intensities, $I_{rel}$, may be obtained as follows:

$$I_{rel} = [(I_{std}/c_{std})\,(v_a/v_{std})^4\,(\mathcal{J}_a/\mathcal{J}_{std})]C \qquad (4)$$

Here the subscript *std* refers to the standard and the subscript *a* refers to the analyte. This equation may be simply expressed as

$$I = K'C \qquad (5)$$

where

$$K' = [(I_{std}/c_{std})\,(\upsilon_a/\upsilon_{std})^4\,(\mathcal{J}_a/\mathcal{J}_{std})] \qquad (6)$$

As mentioned above, internal standards are not always used and, indeed, were not used in two recent studies. In the former, a series of inorganic ions were investigated in the millimole range and correlation coefficients were obtained that were greater than 0.99. The latter dealt with the quantitative determination of benzene, toluene, ethylbenzene, and xylene in water. Detection limits as low as 1.4 ppm were obtained in this study but unfortunately correlation coefficients were not reported.

One of the difficulties that is certainly encountered in the absence of internal standards concerns the $K$ of Eq. (3). Thus, the signal observed at the detector is a function of the laser power as well as the alignment of the sample. This means that the laser power must be carefully monitored and sample positioning must be carefully controlled.

## Applications

### 1. Analysis of $PCl_3$ Reactor Material

On-line analysis of hazardous materials in chemical processes is clearly desirable. $PCl_3$, which is a starting material in the production of phosphate esters, is itself manufactured from $P_4$ and $PCl_5$. Knowledge of the $P_4$ and $PCl_5$ concentrations at various times during the process is critical to safe operation. J.J. Freeman and colleagues used FT-Raman spectroscopy to monitor these $PCl_3$ reactor materials. By coupling an FT-Raman spectrophotometer to the sample window of a reactor with a bifurcated optical bundle, they were able to obtain Raman spectra and thus to quantitatively monitor the concentrations of both of these species.

### 2. Analysis of Aviation Turbine Fuel

Raman spectra obtained with a conventional system (double monochromator, Ar ion laser, and PMT) for aviation fuels give characteristic patterns. These may be used to ascertain the general hydrocarbon make-up of the fuels. As can be seen in Fig. 17, the spectrum of JP-4 is dominated by rather narrow bands, whereas the bands in JP-5 are broad. This indicates that JP-4 has a greater amount of aromatic hydrocarbons and JP-5 contains more aliphatic compounds. Once the peaks are classified as belonging to aromatic (narrow bands at 1386 and 1007 $cm^{-1}$) or aliphatic (broad bands at about 145 $cm^{-1}$) constituents, the associated areas can be ratioed to give a measure of the aliphatic/aromatic character of the mixture. Fuel additives such as antioxidants (phenolic in nature) and icing inhibitors (ethylene glycol derivatives) can be seen readily by virtue of their characteristic Raman bands.

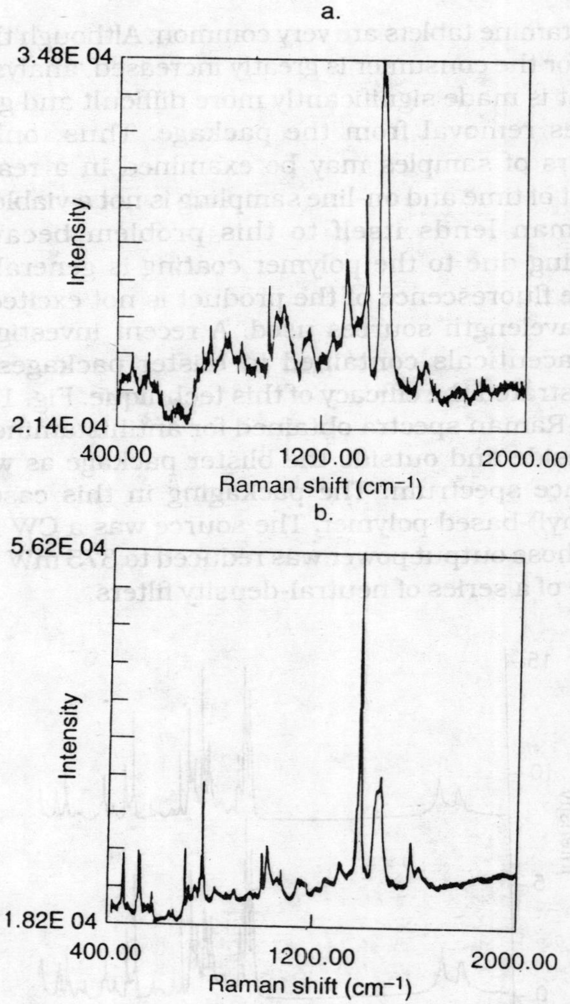


**Fig. 17. Raman spectrum. (a) JP-4 using 514-nm laser excitation and (b) JP-5 using 514-nm laser excitation.**

## 3. Qualitative Analysis of Pharmaceutical Products Inside Packaging

Because of product tempering, pharmaceutical products are now regularly packaged in polymer containers; blister packages such as those used for antihistamine tablets are very common. Although the safety factor for the consumer is greatly increased, analysis of the product is made significantly more difficult and generally requires removal from the package. Thus, only small numbers of samples may be examined in a reasonable amount of time and on-line sampling is not a viable option. FT-Raman lends itself to this problem because the scattering due to the polymer coating is generally weak and the fluorescence of the product is not excited by the long-wavelength sources used. A recent investigation of pharmaceuticals contained in blister packages clearly demonstrated the efficacy of this technique. Fig. 18 shows the FT-Raman spectra obtained for antihistamine tablets both inside and outside the blister package as well as a difference spectrum. The packaging in this case was a poly(vinyl)-based polymer. The source was a CW Nd-YAG laser whose output power was reduced to 375 mW through the use of a series of neutral-density filters.

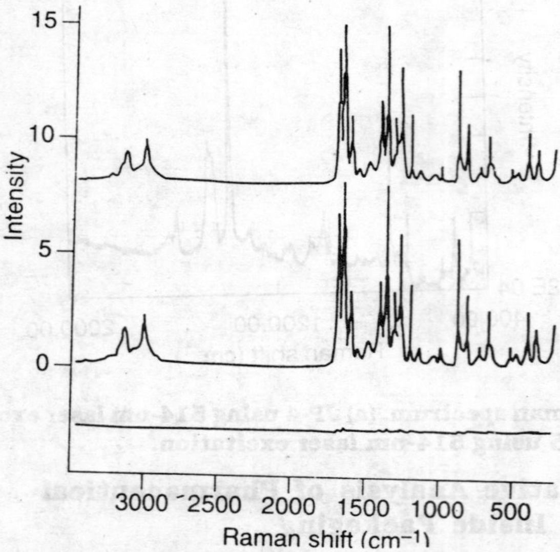


**Fig. 18. FT-Raman spectra of an antihistamine tablet inside its blister pack (top), removed from its blister pack (middle), and difference spectrum (bottom).**

Using this technique, the authors were also able to obtain reasonable spectra from antihistamine microspheres inside gel capsules that were in turn contained in blister packs.

## 4. Remote Raman Sensing

A light scattering technique called differential absorption LIDAR (DIAL) has become a useful technique for measurement of stratospheric molecules. LIDAR is a spectroscopic method whereby the Rayleigh (wavelength of incident light is not changed) scattering from an intense laser beam is collected by a telescope. An eximer laser supplies pulses of radiation and the return radiation is examined with respect to the time elapsed and the intensity in order to produce a concentration-versus-height profile of the analyte. In the DIAL modification, two wavelengths are used, one of which is less strongly absorbed. Typically, lasers emitting at 308 and 353 nm are used for this process. DIAL works well in a clean atmosphere where almost all the scattering is Rayleigh scattering. The scattering of aerosols is ignored or accounted for with a small correction factor. However, introduction of volcanic $SO_2$, such as occurred with the eruption of Mount Pinatubo, causes rapid changes in particle size and aerosol concentration. In order to circumvent this difficulty, McGee and colleagues (22) modified the LIDAR to detect Raman signals from the $N_2$ molecules. The scattering from the $N_2$ is shifted by 2331 cm–1 with respect to the incident wavelength and return depends on the molecular density. Aerosols do not contribute directly to the Raman return and thus backscattering from these is absent. Thus the extinction due to the aerosols can be calculated from the nonabsorbed beam. Using this technique, McGee and Colleagues were able to produce ozone profiles that are valid from 15 to 50 km into the atmosphere.

□□□

